CARPENTER'S
COMPANION

CARPENTER'S COMPANION

Garry Chinn and John Sainsbury

GREENWICH
EDITIONS

A QUANTUM BOOK

This edition printed in 2005 by
GREENWICH EDITIONS
The Chrysalis Building
Bramley Road, London W10 6SP

An imprint of **Chrysalis** Books Group plc

Copyright ©1979 Quarto Publishing Limited

ISBN 0-86288-365-2

QUMWDD

This book was produced by
Quantum Books Ltd
6 Blundell Street
London N7 9BH

Printed in Singapore by
Star Standard Industries Pte Ltd

Contents

CHAPTER ONE

A History of Tools

*Tools, whether of
stone, bronze, iron or steel, are a
result of man's versatility,
his increasing skills through the
centuries, and his mastery of
many skills.*

Tools, whether of stone, bronze, iron or steel, tell the story of man. His need for domestic working and war tools describes his versatility, his increasing skills through the centuries, and his mastery of many crafts. His use of wood is shown through the hundreds of different types which have been scraped, carved, sawn, planed, molded and fashioned to meet his many and varied needs.

Looking at many of the tools of today, you could easily say that they have not changed very much in several thousand years. Nevertheless, they have changed, particularly as man became better at processing raw materials and he began to study the needs of mankind itself.

As far as woodworking tools are concerned, there were three marked areas of development. The Romans made great progress using iron and steel in saw blades and produced jack, smoothing, plough and molding planes, but it was not until after the Dark and Middle Ages that further advances were made. The most significant time was from 1600–1800, when the tenon saw, spokeshave and marking out tools appeared. The screwdriver, all-metal plane, brace, breast drill and auger bits appeared later.

Many tools of the past were made by a craftsman to suit his own needs and, as factories were built to meet the growing demand for crafting tools, they copied the local man-made products. As a result, many of them were quite ornate, yet still functional. They often bore names linking them to their place of origin, and later craftsmen grew accustomed to those particular tools and would accept no other.

In the early part of the Industrial Revolution in Europe, craftsmen and apprentices alike found little time for the leisurely making and embellishment of tools. As a result, the hand-made, individually-designed tool influence declined and made way for more simply designed, purely functional tools.

An illustration of a French cabinet-maker's workshop taken from *Diderot's Encyclopedia*, published in 1760. It shows, amongst other tools, a forerunner of the holdfast.

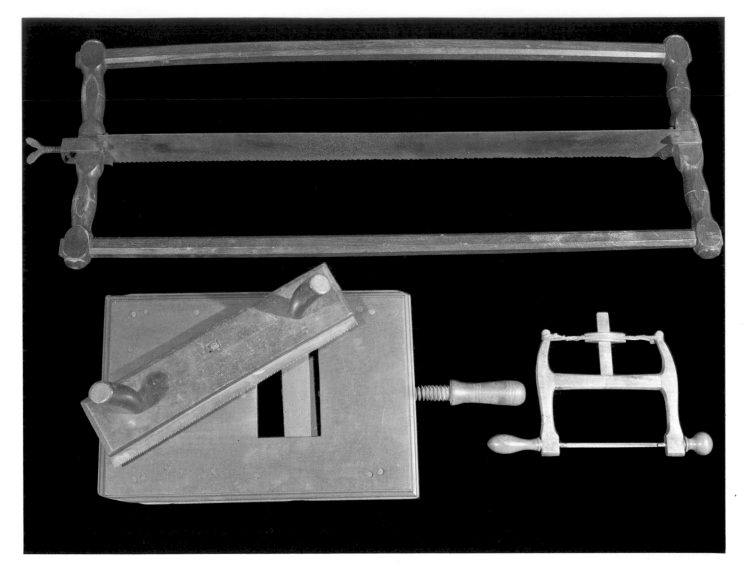

Saws

The saw, like the plane, dates back to the Neolithic or later Stone Age which was well before the discovery of metals. The stone saws were shaped and cut very much as the material itself dictated but the later bronze tool could be shaped and formed. The earliest saws had the teeth inclined towards the handle so that they cut on the pull stroke.

Saws from 100 BC have been found at Glastonbury and one discovered at Cambridge had a blade with 7 points per inch/25mm, all pointing towards the handle. The Romans had frame saws, narrow-bladed hand saws and large cross cut saws with handles at both ends

and they recognized the need for accurate setting of the teeth to right and left.

In the United States Henry Disston did much to influence development, and he began manufacturing in 1844. He made the first crucible cast steel to be used in saws and developed a style and class all his own. Before long, he had captured the American market and was also exporting his saws to Europe. Having developed the skew back saw, he went on to set a standard for others to follow with his rip saw, which had 5½ points per inch/25mm and was 26 inches/650mm long, and his cross cut, which had 7 points per inch/25mm and from 20–26 inches/500–650mm long.

Some beautiful examples of early saws. **Top**, an eighteenth-century veneer saw. **Left**, a French armchair-maker's saw and cramp. **Above**, a small bow saw.

The modern hand saw comes from Holland where, early in the eighteenth century, they designed a pistol-grip handle for greater control and comfort. When the process of rolling wide steel strip was perfected in Sheffield, saws with wide blades were designed. These were quickly followed in the United Kingdom by a handle made from a flat strip of wood with a slot for the fingers and attached to the blade with rivets. By 1750, this pattern was firmly established and in its basic form has survived to the present day.

Boring Tools

Simple copper awls were used by the Egyptians but there is no evidence that they used an auger. The only drills which seem to have been used were bow drills, and these appear in Roman, Greek and Egyptian illustrations. After the Roman period, the breast auger was developed, to be followed later by the brace using iron spoon bits. Many of these spoon bits were fitted with handles and used as augers. The handles were made of various woods, and also bone.

Braces were largely constructed of iron, the bit being held in various ways using a square hole. Perhaps the most beautiful brace ever made, and certainly at that time (1850) the most expensive, was one which is now known as the Ultimatum. This was a patented brace made by William Marples & Sons Ltd, amongst others, in Sheffield, of brass with an ebony rosewood or box infil.

The twisted auger used during the tenth to thirteenth centuries only drilled small holes, and resembled the modern gimlet. Some augers had an eye to take a handle.

As with chisels, boring tools were developed by craftsmen to suit their particular needs, but most were based on a shell principle. A screwed auger which appeared in Smiths Key closely resembles the present-day Scotch auger, apparently invented by a gentleman called Lilley, who lived in Connecticut, although they were being made in the United Kingdom as early as 1812.

The modern auger bit is the result of the early inventive genius of Russell Jennings – an American who in 1855 designed what has been known ever since as the extension bit lip. The solid center bit was designed by Charles Irwin in 1884 – a manufacturing name which is now well-known throughout the world, as well as its native USA.

The modern expansive bit used to cut shallow holes was first patented in 1890 by William Clark of New Haven, USA. A later design by Steers improved the adjustment by using a rack and worm screw.

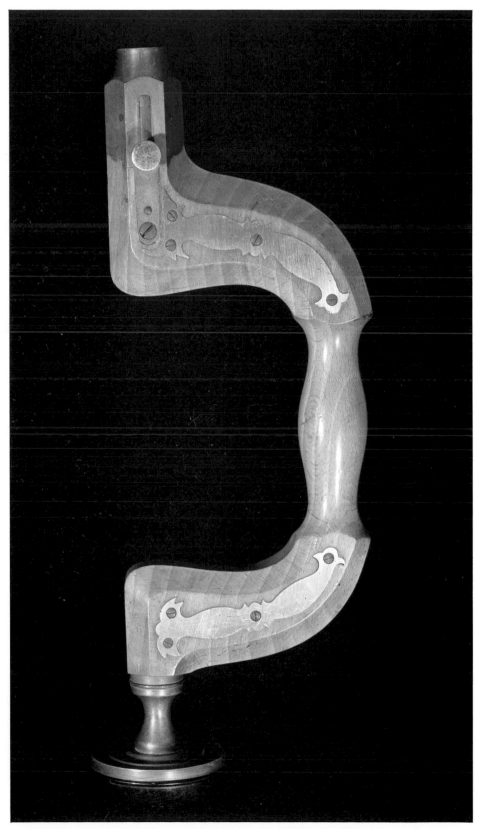

A unique presentation brace by Robert Marples, made of beech with brass inlays.

Planes

In the latter half of the nineteenth century, great progress in the design of factory-made tools was made, credit for which must go to the United States. Many people believe that today's metal plane is a modern design. However, this is wrong as it was Leonard Bailey and his contemporaries who bridged the gap from wood to metal with designs which have survived a century in common use and may well see another century out. The wooden plane of 2000 years ago is still used in many European and other countries today, somewhat modified perhaps, while the metal plane has only been generally accepted in North America, the United Kingdom and the English-speaking countries of the British Commonwealth.

The first iron plane was designed by an American in 1827, who patented a plane with a cast iron stock but no cutter adjustment. TD Worrall of the Lowell Plane and Tool Company followed this development by designing planes with metal superstructures and wooden soles, and a device which tightened the iron in place.

The first really marketable plane, however, came from Bailey in 1858. He introduced a lever cap for the quick withdrawal of the cutting unit and a cutting iron adjustment through a Y-shaped lever (which first worked in the vertical but was later changed to the horizontal), both of which are still used today. Bailey also devised the bent cap iron for perfect functioning of the cutter and introduced the thin cutting iron. Later, he joined with Stanley Rule and Level Company and became superintendent of their plane manufacturing division.

Another Stanley man, Justus Traut, designed the lateral adjusting lever which has also survived to the present day. Various frog adjustments were made by other clever and knowledgeable men so the metal plane, very like the one they knew, has survived a century. Stanley still recognize the work of these men by placing the name Bailey on the front of their bench plane bodies.

The metal combination plane was

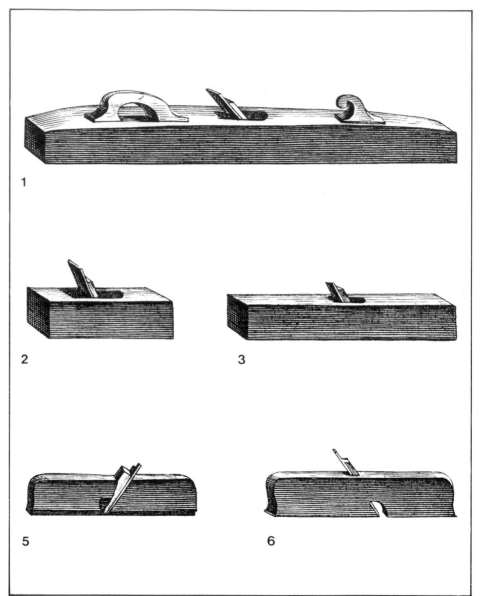

Above left, five planes which were illustrated in *Diderot's Encyclopedia* of 1760. **1** a smoothing plane; **2** a jointer plane; **3–4** rebate planes; **5** a jack plane. **Above right**, part of a late eighteenth-century wall painting in the Hotel de Ville, Paris, showing a trying plane in use. A fifteenth-century illustration, **left**, of a carpenter using an ax. **Right**, a molding plane made about 1725 by the earliest authenticated professional English planemaker, Robert Wooding. This end section shows how the plane cut the wood.

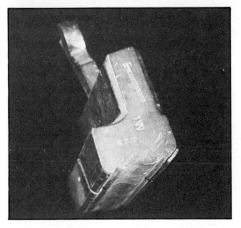

developed at the same time as the bench plane. The first one was patented by Thomas Worrall in 1854 and an iron fillister was made by William S Loughborough in 1859. Charles Miller of Brattleborough most influenced the development of this plane, and his work with Bailey, together with the inventions of Justus Traut, eventually brought about the appearance of the world-renowned Stanley Forty Five. This was further developed into the Fifty Five. The Forty Five is still made by Record Ridgway as the 405. These planes had many cutters and could reproduce almost all the known moldings and other shapes. The planes which came from Stanley were designed to cover almost every known woodworking application.

A few still survive, but anyone who has held and used a Stanley No 444 dovetail dado cutting plane, for example, will have realized and appreciated its remarkable quality.

The steel plane was perfected in the United Kingdom by two companies in an almost identical way – by Stewart Spiers of Ayr and Thomas Norris of London. This plane was developed

1 Craftsman-made left- and right-hand side rebate plane. **2** These highly-decorated shoulder, bullnose and left- and right-hand side rebate planes were made by Preston in the late nineteenth century. **3** Violin-maker's planes. The top one was made by Norris, and the rest by Preston. **4** A screw-stemmed, close-handled solid boxwood plough plane. **Facing page**, a collection of Norris planes – a dovetailed steel panel plane and a smoothing plane, both made about 1920, a steel-soled brass chariot plane and a bullnose plane.

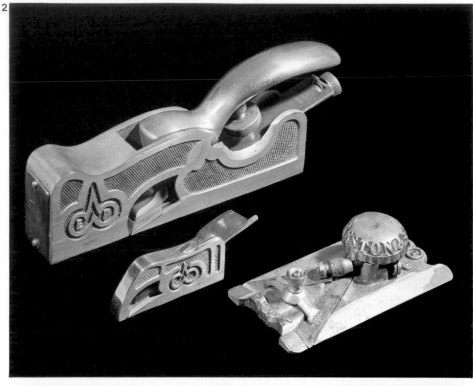

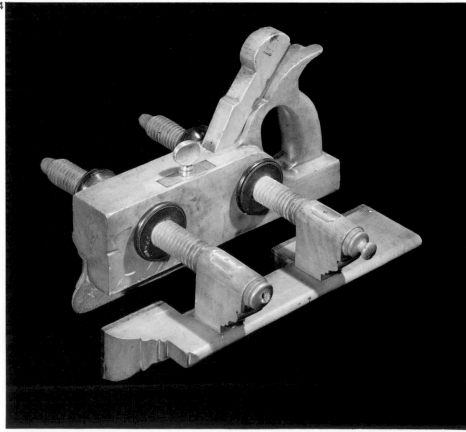

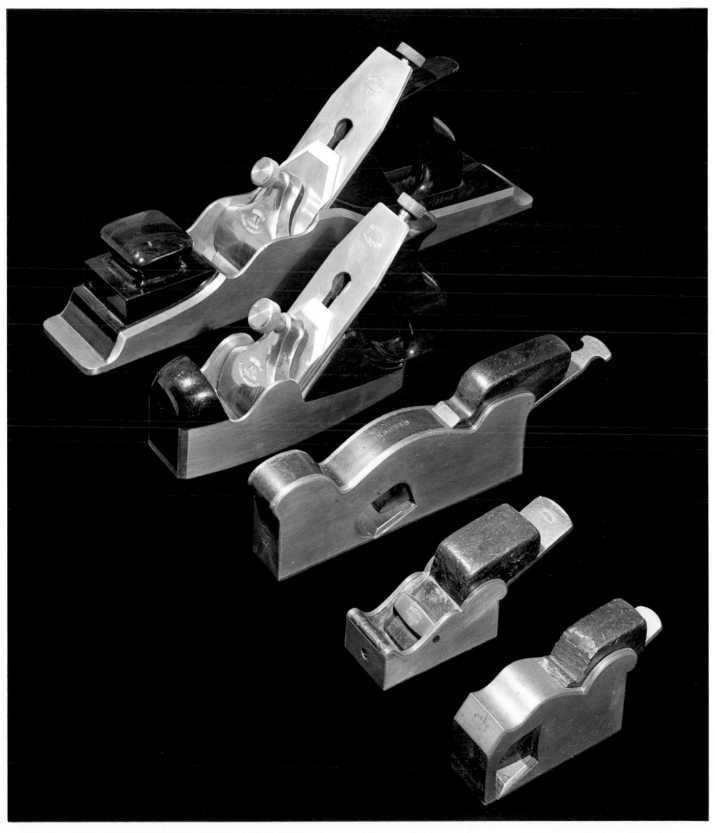

from the early nineteenth-century mitre plane and consisted of a steel box-like body, with sides so finely dovetailed to the sole that often the joints could not be seen. The cutting unit was held in place either with a screw-adjusted lever cap or a wedge, and the body was infilled with timber – usually rosewood. They also made planes using the more conventional grey cast iron or gun metal. The handles were rather differently shaped, and the front handle was usually square with rounded corners and sides.

Smoothing planes either copied the older wooden coffin shape or had parallel sides. They were superb craftsman tools without equal anywhere.

An illustration from a 1909 Stanley catalog showing the Fifty Five plane which was the successor of their innovative Forty Five. The plane is framed by a selection of the moldings and joints which can be worked with it.

Norris was in operation twenty years after Spiers and they provided for the longitudinal and lateral adjustment of the cutter with an invention which appeared in 1913. Both Norris and Spiers planes were expensive but they were prepared to offer unmachined parts for sale, so that a craftsman could make his own.

They also made many special planes, nearly all in the same precise way, with steel and cast iron bodies and rosewood infil. Beautiful steel shoulder, chariot and bullnose planes, often with brass soles, are still seen. Brass violin-maker's planes, made by Spiers, are still available today.

Many special planes were made by Edward Preston, a Birmingham-based company. However, their planes tended to be over-decorated and some were most uncomfortable to hold.

Chisels

The chisel is another tool which dates from the Stone Age. The earliest models didn't have handles, although the Egyptian ones made from bronze and copper did. The close examination of some Egyptian works of art has shown that chisels were used with wooden mallets. The Romans used socketed and tanged chisels – a fresco at Pompeii shows a tanged mortice chisel with a long wooden handle. They were forged from iron bars with a strip of steel at the cutting edge, and sharpened at 25°.

Firmer chisels appeared about 1500 BC and were used with a mallet, but the longer and thinner ones were designed to be pushed. They were handled and had ferrules to prevent splitting. Medieval craftsmen used former chisels, which were flared at the end and used for rough work.

Tool Chests

All these tools, so highly prized, were carefully looked after and often kept in beautifully-made tool chests. These were usually painted on the outside, but the interiors were fitted with drawers made of mahogany or another high-quality hardwood. The lid was almost invariably veneered and inlaid, and sometimes the design incorporated the initials of the owner or a motif of tools. They are extremely valuable and were often handed down from father to son, complete with his quality tools.

This beautiful apprentice tool chest was made in 1858, using a wide variety of woods. The lid is inlaid with ebony, rosewood and ivory, with corner designs in ebony, on a background of birdseye maple surrounded by borders of rosewood, satinwood, quartered rosewood and crotch Spanish mahogany.

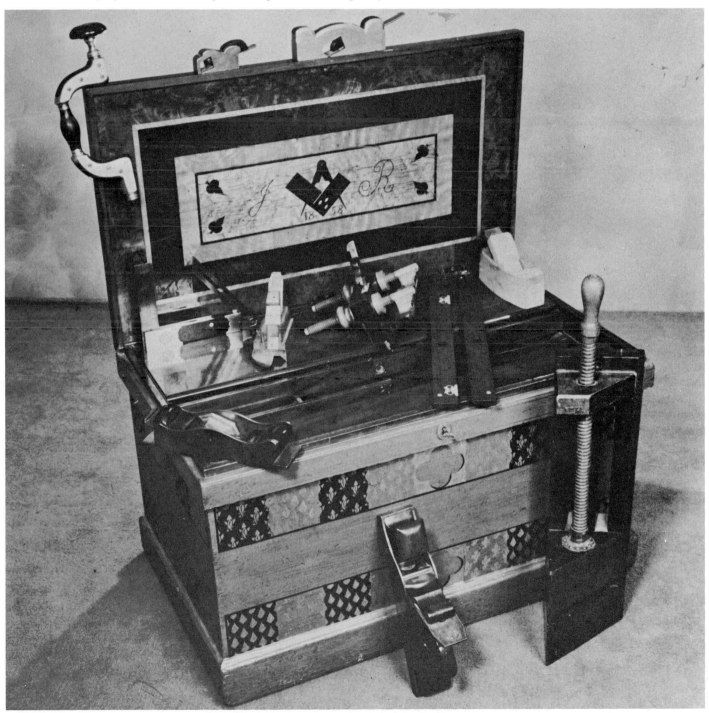

CHAPTER TWO

The Workshop

*To obtain the greatest
enjoyment without frustration,
a workshop is required with its one
essential—the workbench.*

WOODWORKING IS A SKILL built up over many years, the beginner usually starting with a small number of tools. His earliest skills will have been learnt at school in a friendly, relaxed atmosphere, with the relatively sophisticated background of the school and college workshop.

Having acquired a taste for crafting in wood, he may well wish to work in a specialized area, as opposed to one which is more general. This may be brought about by restrictions of space, number of tools, or even their cost. However, to obtain the greatest enjoyment without frustration, he will need a workspace of some kind, with its one essential – a workbench. As his experience grows, so will his need for more tools to advance his work and to save time.

Probably one of the first decisions which has to be made is whether or not the workshop is to be used purely for leisure, or to produce articles for sale. Obviously, thought must also be given to the number of people who may be found working in the shop at any one time. Generally, we can think and plan with the home workshop in mind – expansion of the suggested space and its equipment can naturally follow, should the need arise.

The location of the workshop will depend on a number of factors. Ideally, it should be close to the house, particularly in countries with severe winter conditions, so that an extension of the heating system can be arranged. In North America, most homes have basements, and these are ideal, but care must be taken to prevent or eliminate damp.

Where a separate building is planned, very careful selection of your materials is vital. Galvanized iron sheeting, asbestos or concrete should not be used since these condense badly, bringing about rapid deterioration in tools and materials. A timber or brick building is best. The insulation of both walls and ceiling is important. The floor, preferably wooden, should be treated with preservative but left unpolished – a slippery floor can be dangerous. From time to time, a wooden surface can become highly polished by shavings

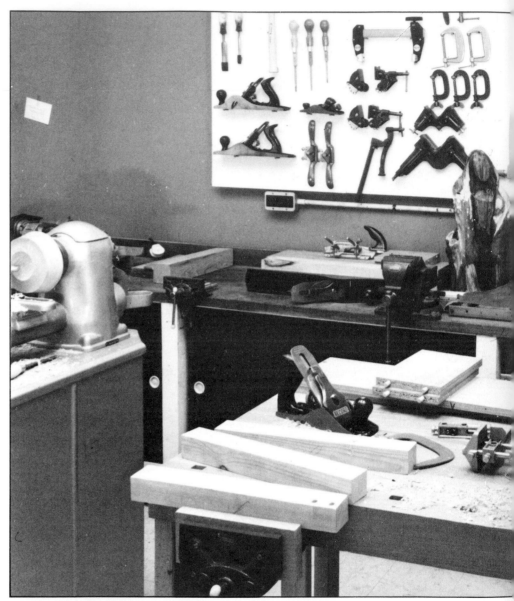

Above, this workshop is very well equipped, with most tools stored on the wall – keeping them tidy and off the working surfaces. **Right,** an easy way to store saws is to hang them on turn-buckles

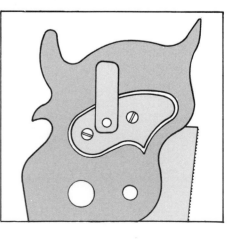

being ground underfoot, but scouring with rough abrasive paper will help to prevent this. A high ceiling will give airiness to the shop, and will be helpful when measuring long timber.

Good natural lighting should be provided, and windows should not be placed near adjacent buildings, as they will obscure the light. If money will allow, then double glazing should be considered. Opening windows are

planer can be equipped with wheels to permit easy storage and conserve space.

Should you choose to set up a workshop in your home, the basement, garage, attic or other spare room will have some of the aforementioned facilities. It will not be difficult to provide more of these, though the elimination of moisture from the garage may prove a problem. The garage is often pressed into use for many craft activities, but it can be restrictive when the layout of tools and equipment is being considered. Nevertheless, it has its advantages when machining large timbers, and often has ideal roof space for timber storage.

The design and construction of a folding workbench, complete with storage space, is vital, if it is considered unsuitable to screw on a simple vice to the kitchen table or worktop.

Planning the Workshop Layout

If at all possible at the outset, draw up a list of the machines and other floor-standing equipment which you will need, even if you buy your tools one or two at a time over a few years. Draw to scale a floor plan, indicate window and door positions, and make from thin card templates to represent each piece of standing equipment. Take each one in turn and place it in the most convenient position, having in mind its use, its lighting needs and the free passage of material through and out of the machine. Shelving and other storage units can follow, and be fitted into convenient spaces. Many woodworkers prefer to have those tools which are most commonly in use, ready to hand and located on pegboard, using purpose-made hooks and pegs. Special planes, housed in their original boxes, will be stored best in cupboards to protect them from dust. Ample space should be left around the machines.

Timber should be stored separately and, when in seasoning, kept in open-ended buildings to ensure a free passage of air. Timber, plywoods, hardboards and multiboards which are ready for use are kept best in a dry situation close to the workshop, but not necessarily inside it.

essential and, in many areas, the addition of fly screens is necessary. A window which can be fully opened will be useful to pass long timbers through when planing or sawing. Generous artificial lighting, using fluorescent tubes of the nonstroboscopic type, should be installed. If a woodturning lathe, morticer and machine drill, band saw or jig saw are installed, they should each be individually lit. Inadequate lighting can be dangerous. As well as individual lighting, a flexible light near the bench for modeling work is vital.

Priority should be given to the positioning of either three- or single-phase electrical outlets. Fusing should be adequate and safety switches provided with each machine. Trailing wires or cables should be avoided wherever possible, and power outlets suspended from the roof for the connection of portable power tools will help to prevent this. A master switch placed conveniently near the door will permit the isolation of all circuits on leaving the shop, and help to eliminate any possibility of short circuiting.

In the small workshop, it will be more convenient if the table saws and

Workbenches and Holding Tools

Having completed the siting of the major items of equipment, a detailed view of the most important piece of the craftsman's equipment – the workbench – is necessary. In assessing costs, due regard must be paid to the bench, because it must be able to meet the needs of the woodcrafts to be carried out on it.

The traditional English bench, with its adjustable bench stop on top and its side vice fitted near the left-hand end, is often quite inadequate. The holding of thin and narrow timber for ploughing, the safe positioning of irregular shapes of timber for sculpting and the gripping of long timbers are just three problems which this bench is incapable of solving safely and properly. There are also many others.

The bench must be of sufficient size and weight, and equipped with the correct holding equipment. Benches on both sides of the Atlantic are fitted with thick timber tops between 2–4 inches/ 50–100mm thick, and sturdy timber legs, although there is a growing fashion for metal legs, probably due to the shortage of suitable wood. Many benches are fitted with a well for storing tools. In Europe, the bench top is made of beech, and in North America it is made of maple. Any hard, long and straight, close-grained timber will do, and the person making his own bench will obviously have to be guided by availability and cost. Attention must be paid to bench height, which obviously varies with the user's own height. Final adjustment to suit individual need can be made by placing stout blocks of wood under the bench legs.

Many manufactured benches use joint and bolt constructions, enabling the user to take up any slack with a spanner, and with the frame ends glued and pinned. Bench tops are usually bolted on, with the bolt heads hidden by wooden plugs.

The European bench is of a different design, with markedly superior holding features. These benches use a wooden tail vice located on the right-hand end of the bench. This vice houses a bench stop, or dog – a steel peg. Holes, equidistant along the bench top, and in line

This beautiful, classic, English workbench shown **far left** is made of beechwood, and held together with coach bolts. This model has a slot along the back edge in which to store tools and a shelf beneath the top to hold pieces of timber and large tools. Both vices have quick-release levers, and the one on the right-hand side is fitted with a tail vice with a sliding dog. However, benches are also available with other vices, such as the ones shown **left.** A rather more unconventional tool is the combination vice, **above,** which rests on, or extends from, any flat surface up to 2½ inches/62mm thick. It will hold flat, round and triangular workpieces.

Making a Portable Workbench

This is a simple folding workbench which can be stored in your kitchen or spare bedroom. It is equipped with a vice, bench stop, tool rest and a bench holdfast. The whole bench can be constructed using standard timber and fittings easily bought from a DIY shop.

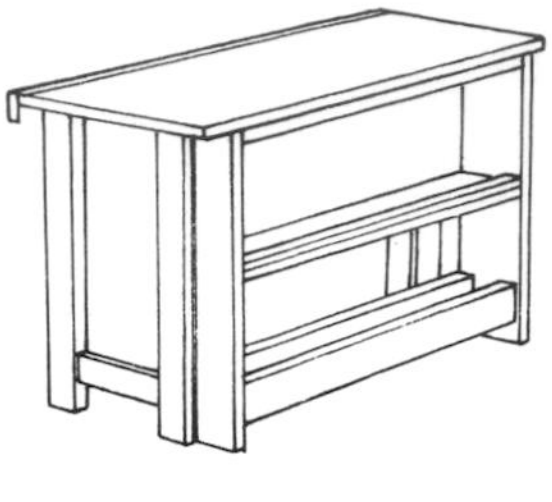

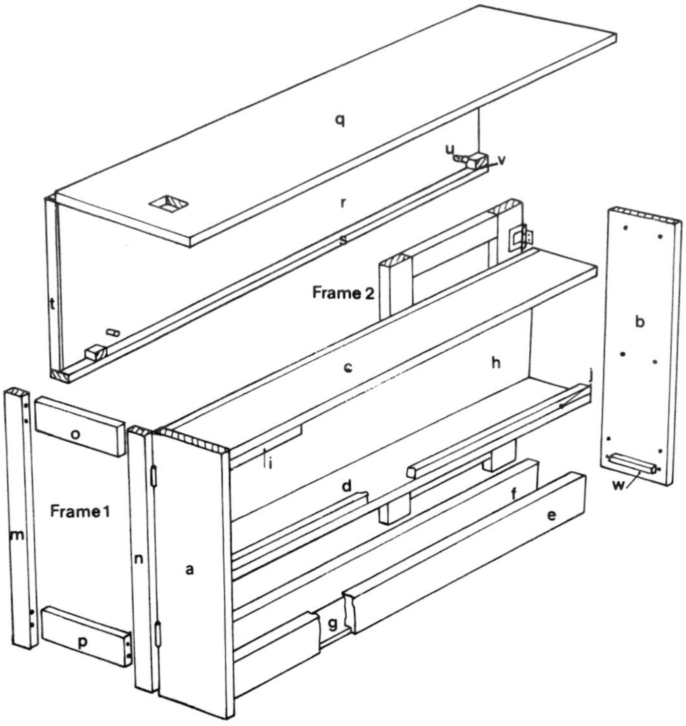

Making the Main Frame

Cut and prepare pieces (a), (b), (c), (d), (e) and (f). If you are using planed timber the only preparation necessary is to cut all the above pieces 30⅜ inches/760mm long. Make sure that all the ends are sawn squarely if they are not to be finally planed.

Set the bush carriers and fences to the sizes shown in diagram 1.

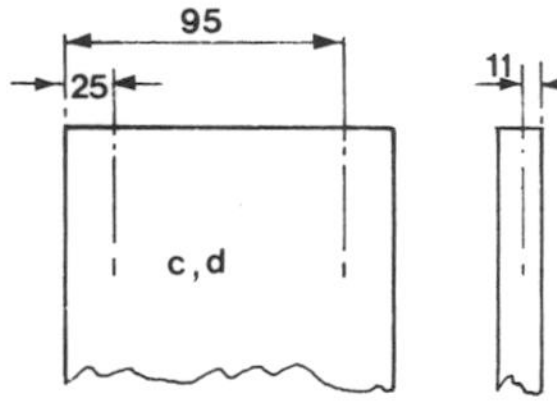

Fig 1

Assemble the jig on the end of piece (c) and drill both holes 1 inch/ 25mm deep. Repeat this process on the other end of the piece and for piece (d).

Fix the adjustable head at its limit on the rods and assemble the jig on the inside and against the end of piece (a). Drill both holes ⅝ inch/ 16mm deep (diagram 2). Repeat on the inside of piece (b).

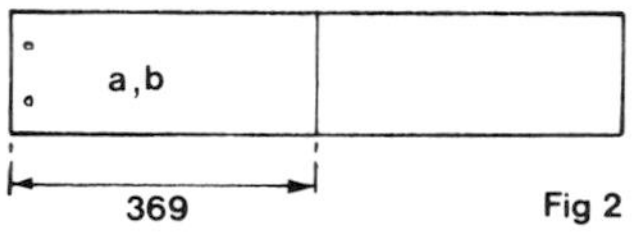

Fig 2

Mark lines squarely across the inside of pieces (a) and (b) as shown in diagram 2.

Remove the fences and locate the jig along the line on piece (a). Drill both holes. Repeat on piece (b).

Reset the bush carrier datum lines at 1 inch/25mm and 2¾ inches/ 70mm and replace the fences at 7/16 inch/11mm.

Assemble the jig on the end of piece (e) and drill both holes. Repeat this on the other end and also on both ends of piece (f).

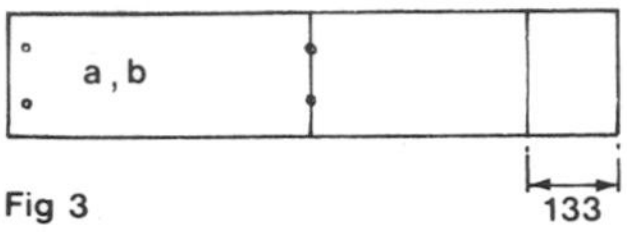

Fig 3

Draw lines on pieces (a) and (b) as shown in diagram 3. This is the center of the top holes to locate the rails of the tool tray.

Remove both the reference head and the adjustable head and attach the doweling jig to piece (a) with a jig clamp, fences against the edge.

Reassemble the jig against the other edge and repeat for the holes to receive the back rail of the tool tray. Repeat both stages on the inside of piece (b).

As drilling for dowels has now been completed, cut 16 pieces of ⅜-inch/9-mm wide dowel 1½ inches/38mm long, and assemble the whole framework dry, to check it. Do not glue up at this stage.

Cut rebates 7/16 inch/11mm wide × 3/16 inch/4mm deep in the bottom edge of pieces (e) and (f) to locate plywood base of the tool tray. Alternatively, a 9/16 inch/13mm square or quadrant strip may be attached to form a rebate.

Making the Legs

As both frames are identical, instructions will not be repeated for the second frame.

Cut all four legs (m) and (n) 30⅜ inches/760mm long and all four rails (o) and (p) 10⅝ inches/265mm long.

Set bush carrier datum lines (diagram 4) and fence to 7/16 inch/ 11mm.

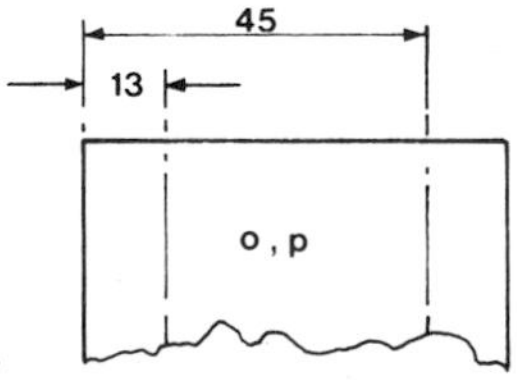

Fig 4

Assemble the jig on the end of piece (o) and drill both holes 1 inch/ 25mm deep. Repeat on the other end. Repeat for (p), and then repeat this procedure on the second frame.

Remove the adjustable head and assemble the jig against the end and on the face edge (x) of the leg (m) and drill both holes.

Repeat on leg (n) and also on both identical members of second frame.

Mark a line across the face edge of all four legs (diagram 5). This is the center line of the top hole of the bottom rail.

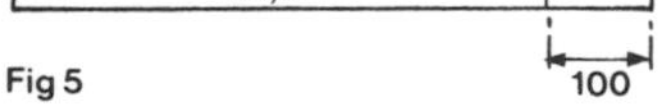

Fig 5

Cutting List

All dimensions are finished sizes. No allowance has been made for waste or cutting. Sizes are given as imperial/metric.

Sundries

2 yards 2 feet/2.4 meters of ⅜-inch/ 9-mm diameter dowel
2 pairs 2½ inch/63mm flush butt hinges
2 pairs 2 inch/50mm back flap hinges
2½ inch/63mm hook and eye
2 dozen ¾ inch/19mm × No 5 countersunk screws
2 dozen ⅝ inch/16mm × No 10 countersunk screws
2 dozen 1¼ inch/32mm × No 8 countersunk screws

Suggested Bench Fittings

No 50 Record Vice
No 169 Record Bench Stop
No 145 Record Bench Holdfast
or equivalents

Main Frame	*Length*	*Width*	*Breadth*	**Material**
2 Legs (a) and (b)	30⅜/760	5⅞/146	⅞/22	Softwood
Top rail (c)	30⅜/760	4¾/120	⅞/22	Softwood
Tool shelf (d)	30⅜/760	4¾/120	⅞/22	Softwood
2 Bottom rails (e) and (f)	30⅜/760	3¾/95	⅞/22	Softwood
Tray bottom (g)	30⅜/760	4 15/16/124	3/16/4	Plywood
Tool panel (h)	30⅜/760	15¼/380	¾/19	Blockboard
Vice mounting block (i)	8/200	4¾/120	¾/19	Blockboard
Strip (j)	30⅜/760	¾/19	¾/19	Softwood
2 Tray supports (w)	4⅛/102	9/16/13	9/16/13	Softwood
Legs				
4 Legs (m) and (n)	30⅜/760	2¼/57	⅞/22	Softwood
4 Rails (o) and (p)	10⅝/265	2¼/57	⅞/22	Softwood
Tops				
Fixed top (q)	36⅝/914	7¾/194	¾/19	Blockboard
Hinged top (r)	36⅝/914	13⅝/340	¾/19	Blockboard
Back edge strip (s)	33⅛/927	1¾/44	⅞/22	Softwood
4 Edging strip (t)	21¼/533	¾/19	¼/6	Hardwood
4 Locating pegs (u)	1¼/32	⅜/9		Dowel
4 Stops (v)	1/25	9/16/13	9/16/13	Softwood
4 Vice cheeks	7¼/180	2¾/70	⅜/9	Softwood or Plywood

Remove both the reference head and the adjustable head and attach the jig to the edge of the leg. Drill both holes. Repeat on other legs.

Having completed all the necessary drilling of leg frames, cut 16 dowels of ⅜-inch/9-mm diameter approximately ⅛ inch/3mm less than the combined depth of both holes: assemble, check and finally glue both frames.

Preparing the Top

It is assumed that all the blockboard will have been purchased and cut to size, so that little preparation will be necessary. If you don't want the core strips of the blockboard to be seen, then apply ¼-inch/6-mm wide edging strip to the end of the top. All measurements are made excluding such edging.

Since maximum strength is required in the top, observe the direction of the core strips when

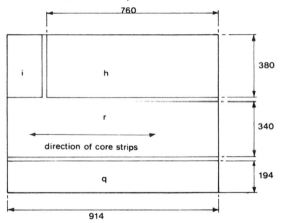

Fig 6

cutting the blockboard as diagram 6.

It should be noted at this stage that if the bench is intended for left-hand use, the mortice for the bench stop should be cut on the right-hand end of piece (q). All other relevant measurements and locations must also be reversed.

Lipping may be added to both ends of the top if wanted to conceal the end grain of the core strips. The measurements in diagram 7 exclude any strip that may be applied.

Mark out and cut a through mortice to receive the bench stop as shown in diagram 7. Most of the waste wood can first be bored out and finally cut with a suitable chisel.

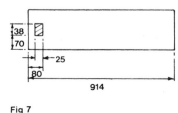

Fig 7

The edge nearest the hinged top must be planed to slightly less than a right angle so that the top can be lifted sufficiently to clear locating dowels when moving the legs into their open position.

Glue and pin a piece of material 1¾ × ⅞ inches/44×22mm to the back edge of the hinged top before finally bringing the two pieces together. The pieces of the top should be hinged together before finally screwing the top to the main framework.

Since there is insufficient space between the back edge of the fixed top and the outside surface of the leg frames to permit a full width of hinge flap to be face fixed, reduce one of the flaps to a two-hole fixing by removing 1 inch/25mm of the flap as shown in diagram 13. Hinges should be placed at each end and at 8-inch/200-mm intervals. Use ⅝ inch/16mm × No 10 screws for the top.

Assembling the Bench

Before screwing the top to the main framework, drill screw clearance holes of 3/16 inch/4mm diameter in piece (c). The holes are positioned so that they do not interfere with either the vice or the holdfast collar fixings (diagram 8). These holes should be countersunk on the underside. 1¼ inch/32mm × No 8 countersunk steel screws should be used to screw the top in place.

Vice mounting block (i) 8 × 4¾ × ¾ inches/200 × 120 × 19mm blockboard can be screwed to the underside of piece (c) having been drilled to take 1¼ inch/32mm × No 8 screws (diagram 9).

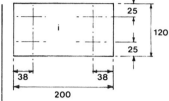

Fig 9

Cut tool panel (h), trim and fix in position. It is finally held with 1¼ inch/32mm × No 8 screws, having been drilled and countersunk as shown in diagram 10.

Having cleaned up the leg frames,

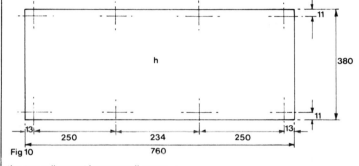

Fig 10

the top rails must have small areas removed in order to clear the hinge knuckles which are face-fixed on the underside of the top, diagram 11.

Leg frames may be hinged to the back edge of the main frame using ¾ inch/19mm × No 5 countersunk steel screws. Use flush butt hinges so they are flat mounted, as shown in diagram 12.

Both legs which are to take hinges should be shortened by 9/16 inch/13mm to prevent shavings etc being trapped when the legs are moved to the closed position. When screwing up the hinges, turn the bench upside down.

Raise the hinged top into its open position and move the legs also into their open position. Bore a ⅜-inch/9-mm diameter hole through the top and into the top of the leg. Repeat for the other locating dowel.

Close up the legs and glue into holes in the bench top a piece of ⅜-inch/9-mm dowel, 1¼ inches/32mm long to finish flush with the surface of the top.

Thin pads of rubber glued to the bottom of the legs will prevent any tendency for the bench to slide if

used on a smooth floor surface. A small pad glued to the top edge of the rail (e) will act as a foot pad.

When fixing the bench holdfast it is recommended that the first collar is fixed at 7⅝ inches/190mm from the right-hand end of the bench and the second 15³/₁₆ inches/380mm from the first. Both centers should be located at 3⅝ inches/90mm from the front of the bench.

A 2½-inch/68-mm long lightweight hook screwed to the bottom edge of the tool panel and an eye screwed to the outside of the back edge strip (s) provides a means of securing the moving parts of the

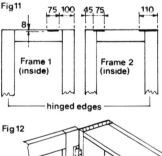

Fig 11

Fig 12

13mm cut off end of leg

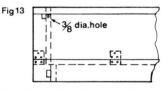

Fig 13

⅜ dia. hole

bench together when carrying.

In order to protect both tools and vice from unnecessary damage, suitable wooden cheeks 7¼ × 2¾ × ⅜ or 9/16 inches/180 × 70 × 9 or 13mm should be screwed to the vice jaws before screwing the vice to piece (i). The top edges of the cheeks may be planed flush with the bench top after attaching the vice.

No suggestions have been given for tool panel or tool shelf layouts as individual requirements may vary.

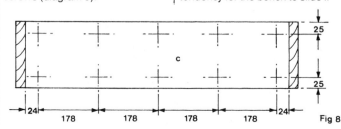

Fig 8

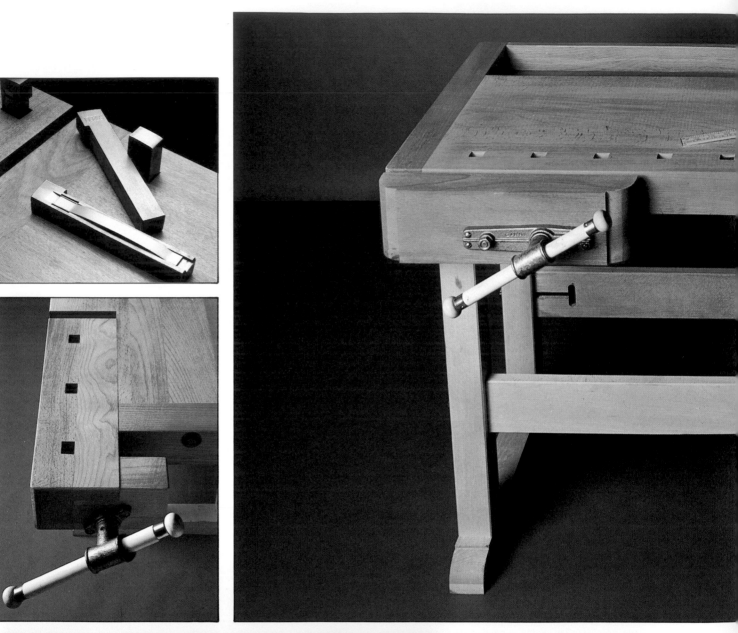

A small European workbench, with all the characteristic features of its type. Regularly-spaced holes ranged across the bench top accept dogs, **top left,** which can therefore be placed in any position along the top. This tail vice, **center left,** is very useful for holding wood which is to be shaped and carved. **Bottom left,** a piece of carving is held between a tail vice and a bench dog. A larger tail vice, **near right,** is able to hold very large pieces of wood. A vice clamp with a variable pressure pad can be fitted into dog holes on the bench top or used vertically, **far right.**

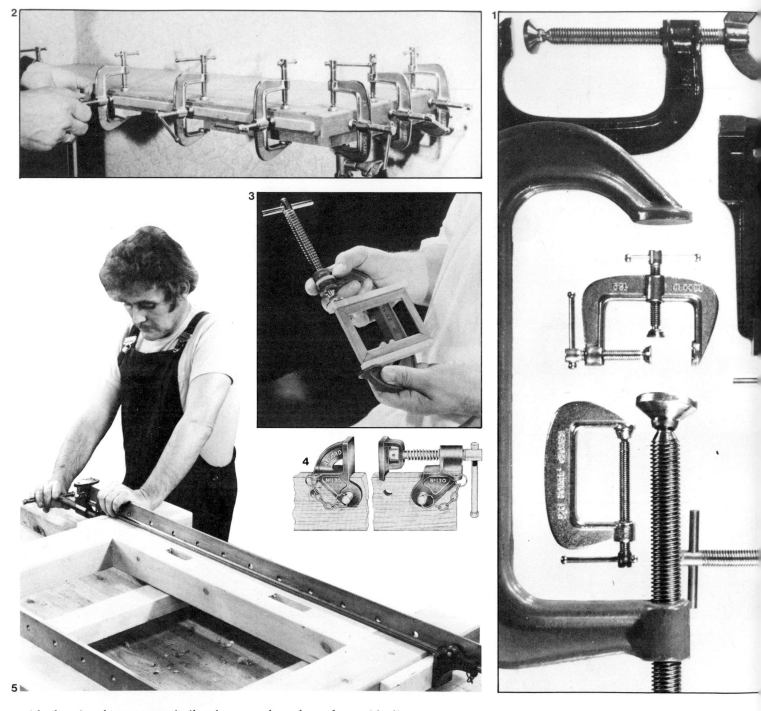

with the vice dog, accept similar dogs. Thus, any length of timber can be held between the dogs, using the vice to give rapid tightening. These vices use metal screws, the dogs being of wood, plastic or metal. The side vice is generally made of metal, and is available in plain screw, or alternatively, is fitted with a quick-release lever for rapid adjustment.

The holding of timber for carving or sculpting will require the addition of a bench holdfast or carver's screw. Both these are fitted into a hole in the bench top. The holdfast consists of a shaft, with a lever arm pivoted at the top actuated by a screw. The lever arm has a swivel shoe which is placed on the work, and through which pressure is applied. The manufacturer supplies two metal collars which can be inserted into the bench top, just below the surface, to house the holdfast and save wear and tear of the bench top itself. It is also useful to have a collar inserted near the

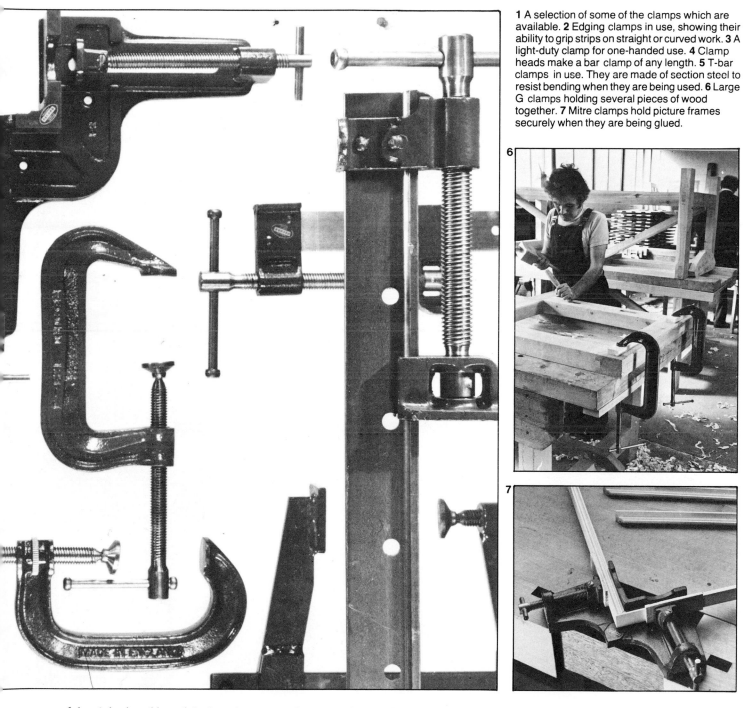

1 A selection of some of the clamps which are available. 2 Edging clamps in use, showing their ability to grip strips on straight or curved work. 3 A light-duty clamp for one-handed use. 4 Clamp heads make a bar clamp of any length. 5 T-bar clamps in use. They are made of section steel to resist bending when they are being used. 6 Large G clamps holding several pieces of wood together. 7 Mitre clamps hold picture frames securely when they are being glued.

top of the right-hand leg of the bench, to allow additional support for long timbers held in the side vice.

The woodcarver's screw, designed to hold wood when it is being carved on each side, is screwed into the underside of the workpiece and passed through the hole in the bench. A wing nut is applied to the screw from below to secure it.

Most individual needs can be met with these traditional tools, but the woodworker will find himself improvising from time to time to solve a particular problem. G clamps (C clamps) used with the bench top or vice, often solve the problem.

Where space permits, a side bench is useful. Boxed oilstones need a home and, since they are constantly in use, a location on a purpose-designed side bench will avoid frustration and save time. Small electric tools such as the grinder and drill can be housed in the same way. These benches need not have

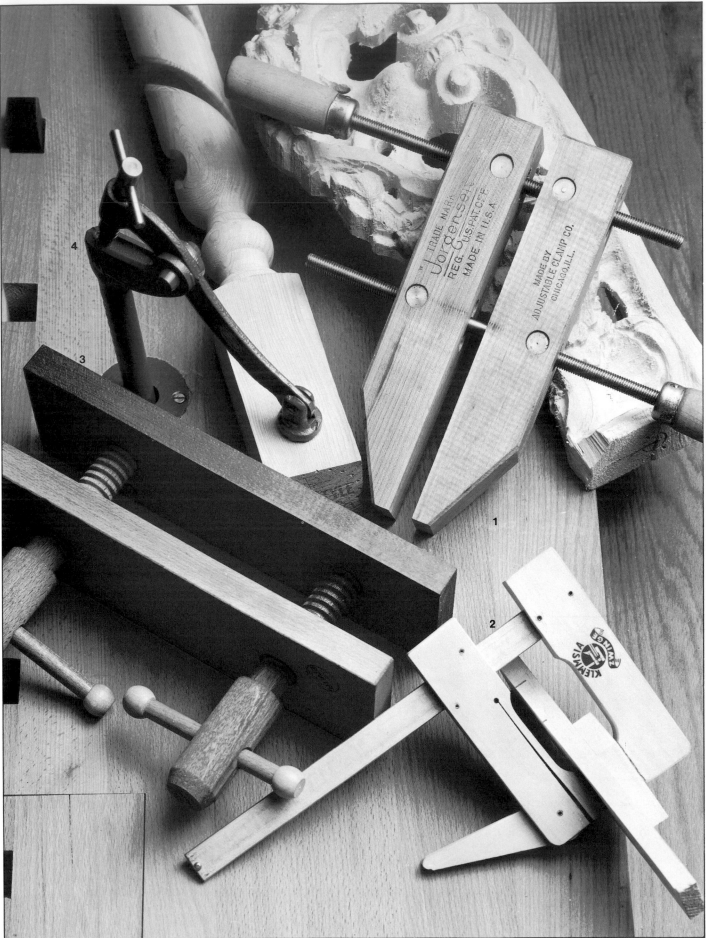

heavy tops, and any available timber can be used.

The sawing of planks and sheets can only be carried out satisfactorily using sawing trestles. These, if made in the traditional way, have splayed legs and need storage space. There are, however, a number of patented devices, made of metal, requiring 3×2 inch/75×50mm or 4×2 inch/100×50mm timber legs which are easily removable and will require less storage space. Trestles made with hinged tops also take up less space.

A bench which doubles as a trestle is the ever-popular Black & Decker Workmate. This is available in several heights, and its top serves also as a vice. It can be folded flat and hung on a wall – the ideal solution for the woodworker with a space problem.

Select your bench with care, and think ahead if possible, to forecast the problems it may need to solve. Consider it too as a partial solution to your storage problem, and remember that you may regret for all-time a poor-quality or inferior selection.

Apart from the traditional clamp, there are clamps for lighter work, some of which are illustrated on these pages. **Left: 1** A wooden hand screw with jaws that close at a variety of angles. **2** A light clamp used for veneering or light gluing but which can also exert great pressure. **3** Beech bench vice suitable for light work. **4** A bench holdfast which can be clamped to any workbench and is especially useful in low-relief carving. **This page: Above,** the speed clamp is a new design in bar clamps, with both head and tail sliding along the bar; **above right,** two fixtures fit onto any ¾-inch/18-mm standard threaded pipe to make a firm clamp which works like a C clamp but is much faster to open and close; **far right,** small, general-purpose spring clamps; **near right,** lightweight sliding head clamps.

Choosing Your Tools

Obviously, the choice of tools will depend upon the particular craft, but the selection of the right tools of the right quality is fundamental. The quality of the tool, with regard to design and material, must over-ride all other considerations. A certain number of hand tools, such as saws, planes and chisels, are essential to the basic kit. Look for those carrying well-known brand names, which invariably carry a maker's guarantee. One or two portable power tools, together with stationary machines, may be useful additions, but should be selected with their likely use in mind. Where weighty or bulky timbers are to be used, it may be necessary to give priority to portable power.

The exploration of timber through hand tools must be the path that the true craftsman will follow. It is only in this way that he can truly discover timber; probably the most beautiful of the raw materials man can fashion. It is only possible to appreciate the problems of a machine in cutting and shaping wood if one has carried out the same work first using hand tools. This is the answer to those people who tend to say, often quite strongly, that the use of hand tools will gradually decrease until power tools take over completely. This may well apply to large areas of industry but can never apply to the serious home craftsman, the craftsman earning his living in a true craft or the student in college or university.

Quality work will almost invariably spring from a sound knowledge of the

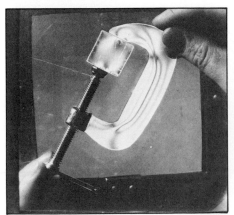

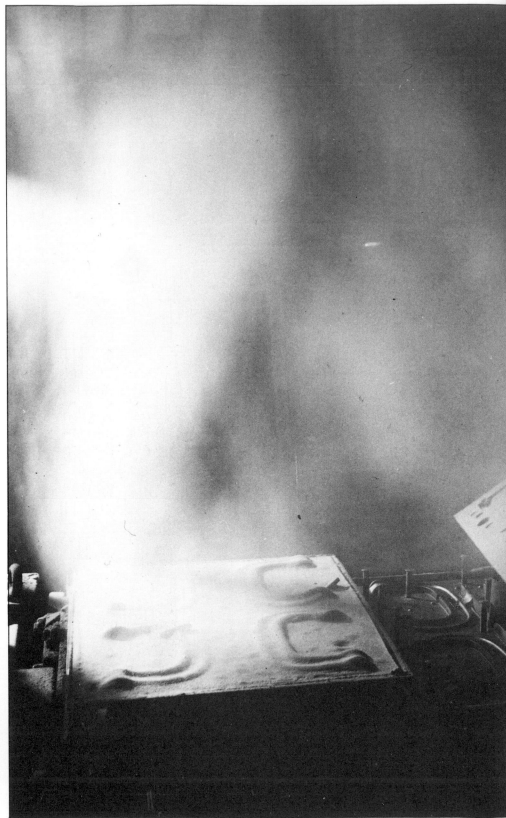

Tools that will last are expensive, as they are made from high-quality materials. New designs are also rigorously tested before being put into production. **Opposite page,** a plastic model of a G clamp is loaded in beam of light and observed through a polarizing screen to test its strength. A shell mold for G clamps is cast, **left. Below,** try squares are meticulously tested for square before leaving the factory. A wood-boring tool is forged, **below middle.** Chisel handles are turned from boxwood logs, **bottom.**

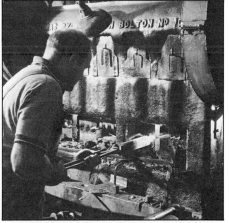

use and function of a number of hand tools. Where timber comes in bulk, a rip saw together with a cross cut saw will cut adequately, but a fine-toothed panel saw will be essential for cutting plywood panels. However, a hand power saw can reduce the hard work and save considerable time, particularly if timber is to be cut from large stock. No matter of how fine a quality the power plane or jointer may be, it cannot produce the perfect surface. This will only come with practice in using the metal jack, or fore, plane. The cleaning up of work after gluing will require the use of a metal smoothing plane.

Other edge tools, such as chisels in their various forms, are essential. There is a vast difference in the quality of the many brand names available. The discerning woodworker will want the chisel to handle well, and be sure that the steel is of the right type to maintain an edge. The wood used in the handle must be selected with great care. In the United Kingdom, the traditional timber used in quality chisels is European Box (*Buxus sempervirens*). Box is extremely dense and close-grained, and therefore resists splitting. The contemporary plastic-handled chisels, using cellulose acetate butyrate or polypropylene, are also splitproof, but many craftsmen dislike using them. They haven't the feel of wood and can also be slippery to hold. The handle, no matter of what material, must be designed to fit the hand of the woodworker. Its shape must take into account the various cuts it makes when held in the right or left hand. A good handle attached to any tool must give a positive grip, yet feel comfortable. Selection must, therefore, be made with this in mind, and a higher price paid, if necessary. Good tools, lasting a lifetime, will give continual satisfaction in use.

In addition, a number of essential tools, such as a mallet and hammer, a drill, a number of screwdrivers, a joiner's or combination square and some clamps will be needed. Several portable power tools, including a jig saw, power drill and circular saw, can follow later. Detailed discussion later in the book suggests when and what to buy.

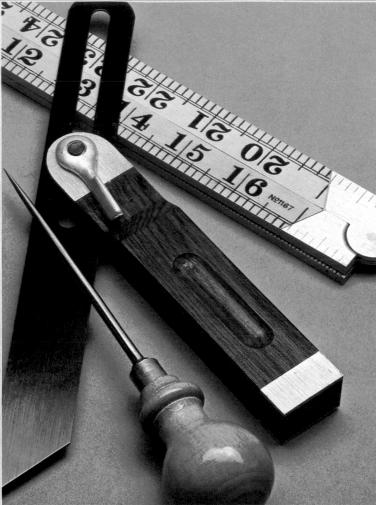

Measuring and Marking

*After selecting and preparing
timber, measuring and
marking tools will be needed
to mark lines for joints and
sawing, and for testing
for true.*

AFTER DESIGNING AND PLANNING a particular project, the timber must be selected and prepared. The preparation may entail sawing and planing, although the extent of the latter may be reduced by the use of man-made boards or pre-planed material. From the outset, however, measuring and marking tools will be needed, not only for measuring and marking lines for joints and sawing, but also for testing for true.

Usually, the woodworker will have ready a detailed drawing or sketch, with measurements added, from which to work. A cutting list is useful, and will ease the work of timber preparation. The transfering of drawing detail to the timber may often appear tedious, but it is a vital stage and one which must not be carried out hurriedly or carelessly. The accuracy of each mark made will depend upon that of a previously-made one, otherwise an accumulation of errors will result in a waste of time and materials. All joints must be marked out accurately and checked to ensure a perfect fit before sawing.

Accurate, well-made tools are vital to ensure perfection in the lay-out stage, and they will always be needed – even in the smallest project. Expensive though they may be, they will pay dividends in the long run.

Rules and Tapes

On both sides of the Atlantic, the traditional measuring tool for the average woodworker has been the 2-foot folding rule, although the joiner and carpenter has favored the 3-foot folding rule. Since the onset of the metric system, both of these are gradually being superseded by the metric rule, in a 4-fold style. The foot rules usually had regular markings as small as ¹/₁₆ inch completely along both edges of one side, with ¹/₈ inch markings on the reverse.

A good folding rule will be constructed of boxwood, with brass fittings. Boxwood was chosen because of its close grain and hard-wearing qualities. However, it is becoming very difficult to obtain, and substitute boxwoods are being used. A number of manufacturers are now using plastics,

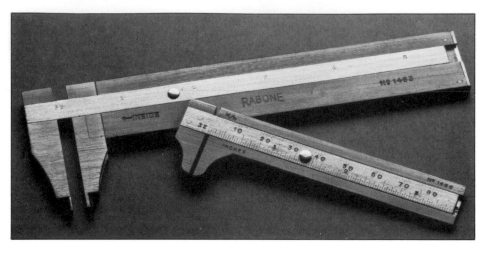

the most popular one being makrolon, with stainless steel fittings.

In countries where Imperial measurements are being phased out, a combination of Imperial and metric can still be found. Some tradesmen, particularly in North America, require a rule of greater length than that provided by the four-fold rule. The zig-zag rule meets this requirement.

It was made in boxwood originally, but several steel versions are now available. A variation of the wooden zig-zag, called an extension rule, has a slide inserted on one arm to take the inside measurements of frameworks and other constructions.

Perhaps the most versatile of the measuring tools is the flexible tape. These tapes were made of linen once, and cased in leather, but the contemporary type is generally of flexible steel in a metal case. They are ideal for measuring in the flat or in the curve, since they can be wrapped around the object. Measuring out using a steel tape often gives more accurate results than when a thick wooden rule is used, because the tape is flatter and lies closer to the object being measured. The majority of tapes are divided into inches down to ¹/₈ inch on one side and in meters down to millimeters on the other. These rules should be handled with care, and a spot of oil or a rub down with an oiled cloth will keep them in good condition.

The two-fold steel rule is popular also amongst tradesmen, and is available in

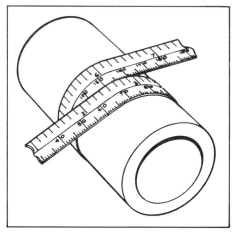

Caliper gauges are finely graduated; the larger gauge **above** can take both inside and outside measurements. **Below:** Measuring the circumference of a tube using a flexible tape.

both Imperial and metric measurements.

One-piece steel rules can be useful sometimes. However, they are designed for the engineer who needs to take very small measurements. They are obtainable in lengths of 6, 12, 24 and 36 inches, as well as 300mm and 1m.

Calipers

Although not the most important of the measuring tools, there are times when calipers are indispensable, such as when taking inside or curved measurements.

Calipers are usually made from high-quality steel, the legs being held firmly in any position by their hinge pin. This type can often be upset by the slightest pressure so, if they are to be used a great deal, it is best to buy a pair of spring-divided calipers.

The spring-divided caliper arms are set up on a bow spring, and are adjusted to size by a nut and screw.

Both plain and spring-divided calipers are available as inside and outside calipers in maximum widths of 4 inches/100mm, 6 inches/150mm and 8 inches/200mm. The inside calipers have straight legs with tiny feet at the extremities, and the outside calipers have curved legs to clear the work.

For the woodturner and other craftsmen, the double caliper is most useful. This works as both an outside and an inside caliper and is most useful in measuring lathe work (eg making boxes with lids), to ensure a close fit.

Inside and outside measurements can also be taken using a caliper gauge. These are manufactured in boxwood and brass, as well as in solid brass. Graduated finely, they are particularly useful when taking readings on small articles (eg the fitting of a ferrule on a chisel handle). They consist of a boxwood body and a solid brass slide, which can be moved by pressure on a small stud. Large versions are also available. These are made of steel and are of engineer's quality.

The measuring and transfering of angles calls for the use of a protractor. Those used by the draughtsman are unsuitable, usually being made of plastic, with measurements which are difficult to read. The most suitable tool is the craftsman's protractor, which has a square head, unlike that of the drawingboard protractor, and an arm which is held in any position by a knurled nut. Graduations are in 1° intervals, from both right to left and left to right. They are usually made in high-quality steel, and can be used in marking out timber, setting up machinery and in a host of other odd jobs.

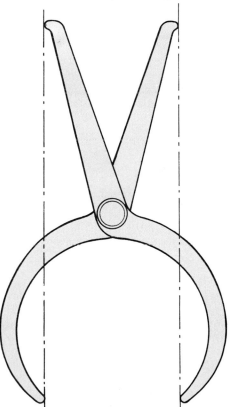

The double caliper **above** has a pair of straight legs for taking inside measurements, and curved legs for outside measurements. Both pairs of legs always measure the same, so that any inside measurement taken with the straight legs can be immediately used as an outside measurement with the curved legs. **Left: 1** A small engineer's square for precision joinery. **2** A craftsman's protractor with a 0–180° scale in both directions. **3** A spring-divided caliper for measuring inside dimensions. **4** Dividers have hardened points and are used for scribing arcs and circles. **5** A spring-divided outside caliper with curved legs.

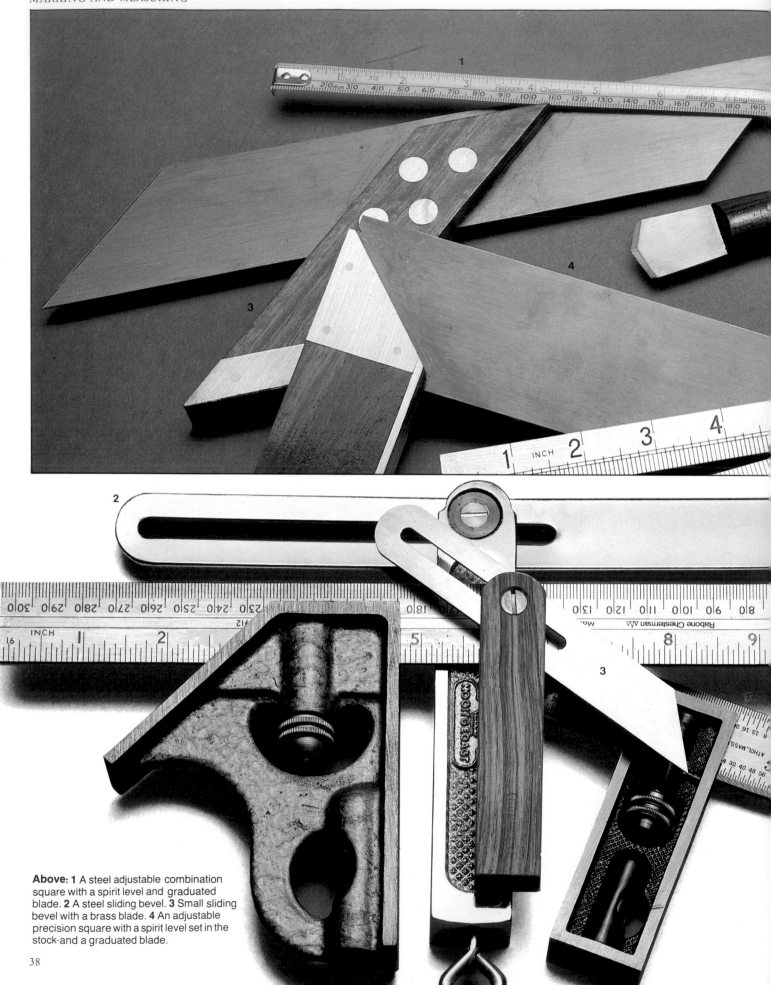

Above: 1 A steel adjustable combination square with a spirit level and graduated blade. **2** A steel sliding bevel. **3** Small sliding bevel with a brass blade. **4** An adjustable precision square with a spirit level set in the stock·and a graduated blade.

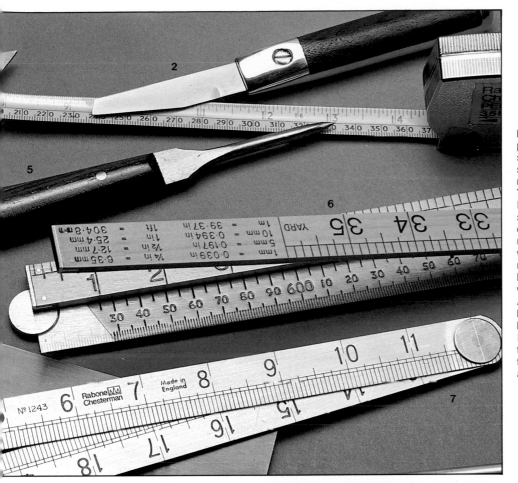

Left: 1 A flexible steel tape rule. **2** A general-purpose knife with replaceable blade. **3** 45° square for marking mitres. **4** A combination try square and mitre square. **5** Combined awl and marking knife. **6** Three-foot folding boxwood rule. **7** Two-foot brass single-fold rule.

Below, a collection of squares. **1** A deluxe square with rosewood handle. Both inside and outside edges are fitted with dovetailed slotted milled brass facings. **2** A very rigid try square, with a handle faced with solid brass. **3** A model-maker's square with a 2-inch/50-mm brass blade. **4** An African satinwood-bladed square with a precision-milled African Padauk handle. As the blade is 15 inches/375mm long, it would be too heavy if made of metal. **5** A solid brass-bladed square with a Brazilian tulipwood handle. It is very useful for cabinetwork. **6** A shock-proof try square with a blade that runs the full length of the handle. It is impossible to knock it accidentally out of square.

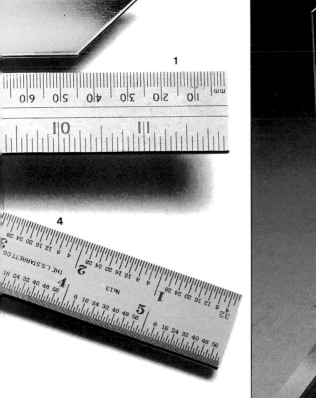

Squaring Tools

The right angle, or 90° angle, is one which is vital to most constructional work. In fact, it is impossible to think of a woodworking construction which does not need the square. The jointing of timber almost invariably requires great accuracy in fitting shoulders, and shelves and doors must also fit exactly.

The try square is the basic squaring tool required for marking out and testing angles. It consists of a blade of steel accurately set at 90° in a wooden (or metal) stock. In the best-quality squares, wooden stocks are made of rosewood, and are faced with brass to prevent wear. They are usually available in 6-, 9- or 12-inch/150-, 225- or 300-mm blades, and a number are graduated in inches and millimeters. A good-quality try square has been precision-tested for accuracy, and is sturdily riveted to resist shock. The best type has an L-shaped blade, which extends the length of the hardwood stock.

On a number of squares the top of the stock is cut at 45°, so that mitres can be marked, eliminating the need for a separate tool.

A variation on the square is the mitre square. Designed to mark and test mitres, the angle of 45° is fixed, with the blade fixed to a hardwood stock.

Mitre squares are also available in steel, but these are not widely used by woodworkers, who prefer to work with wooden ones. In recent years, the versatile combination square has become popular. Made with a steel or alloy stock, the steel blade slides along the stock, which is faced at 90° and 45°. The blade is instantly secured in any position. Generally it is marked in inches and millimeters. This tool can serve as a square, mitre square, depth gauge and, in conjunction with a pencil, it can be used as a pencil marking gauge. A number are fitted with a small vial for use as a spirit level, and a small scribing pin for marking out. The woodworker's combination square should not be confused with the engineer's version, which is a tool of great precision and extremely expensive.

For the marking of any angle, and particularly the dovetail, the wood-

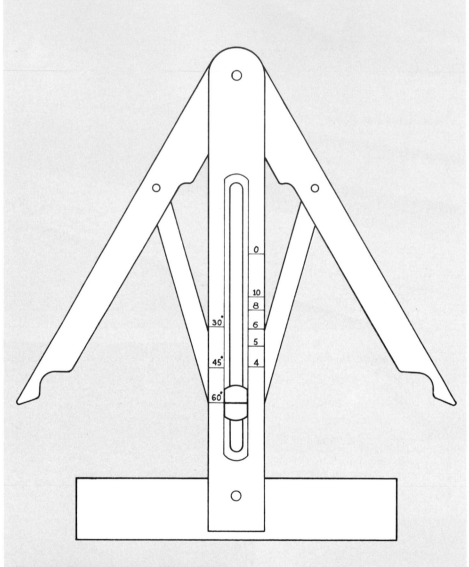

Many-sided figures can be constructed using the angle divider. The body is marked for 4, 5, 6, 8 and 10 sides and 30°, 45°, and 60° angles.

worker will need a sliding bevel. This tool has a hardwood stock and a sliding, slotted blade, which can be secured at any angle by either a slotted or winged screw. It is usual to set up the required angle on the edge of a board and align the bevel to that, or to set it using a protractor.

In built-up work, either on the bench or lathe, the setting up of angles other than 90° or 45° presents a problem. The angle can be struck from a protractor but this can often prove difficult and give unsatisfactory results. Often too, the woodworker has to transfer an angle from one piece of timber to another. The angle divider is the answer, the steel arms and body can be locked into any position and set to accurately marked graduations. The graduations give settings from 4-, 5-, 6-, 8- and 10-sided figures, as well as 30°, 45° and 60° bisected angles. It is not an essential tool for the average kit, but nevertheless a

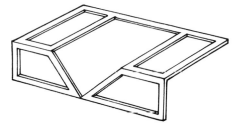

A mitre template for marking out 45° and 90° angles.

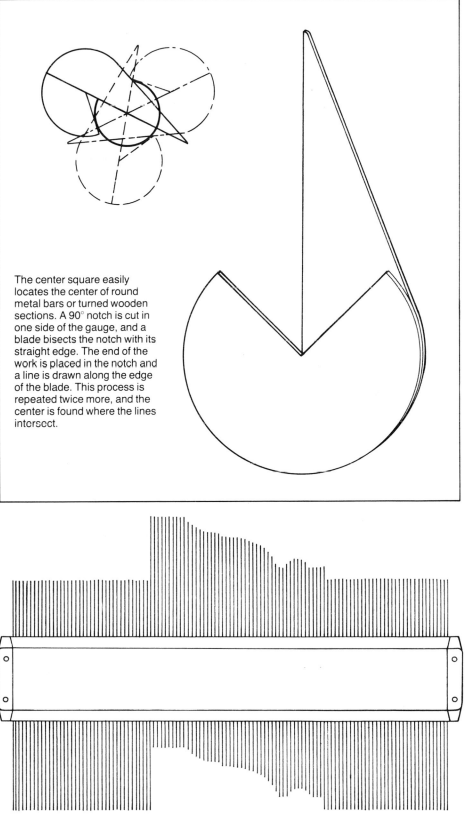

The center square easily locates the center of round metal bars or turned wooden sections. A 90° notch is cut in one side of the gauge, and a blade bisects the notch with its straight edge. The end of the work is placed in the notch and a line is drawn along the edge of the blade. This process is repeated twice more, and the center is found where the lines intersect.

time-saver which will greatly improve accuracy in setting out and testing.

A more unusual square, particularly for centering circles and circular work, is the center square. The woodturner will find this tool most useful for locating accurately centers for spindle turning. There are a number of styles, from the very simple, to one which combines a protractor and is used to measure angles.

The mitre template is an ideal tool for marking out at 45° and 90°. Usually 6 inches/150mm long, it is accurately ground, and extremely useful for marking out battens and smaller timbers.

To test the flatness of boards, and for marking lines on plywood and other large surfaces, the craftsman will need a number of straight edges. He usually makes these from some good, long and straight, well-seasoned hardwood. Mahogany or beech are ideal, when suitably protected with polish. If the board is to be cut, a steel straight edge is needed. One with one edge chamfered would be best – this is particularly important for the man cutting veneers.

Occasionally, a spirit level may be needed, to check either horizontal or vertical levels. It is most useful whenever a cabinet has to be fitted to a wall, or a door lining has to be checked for true. They are available in wood or metal, often with a sophisticated combination of two-way vials. The woodworker needs a simple spirit level, unlike the builder or bricklayer. The modern tool has a plastic vial, rather than glass, and is usually filled with alcohol.

A pattern can be made of any complex shape using a profile gauge. It is composed of sliding steel needles which, when pressed against an object, will faithfully reproduce its shape.

Marking Out Tools

The marking out of joints, almost invariably, is done with the pencil but, whenever timber is to be cut, lines must be marked with a knife. This severs the top fibers of the timber and leaves a sharp corner, in which to insert the chisel or saw.

A finely-pointed pencil is better than the traditional carpenter's oval pencil, which fails to give critical markings. A sharp knife, preferably ground on one side only, is essential for achieving the best results. A number of styles are available, with wooden and plastic handles.

An improved marking tool is the combined awl and marking knife, with a rosewood handle. The knife permits both right- and left-hand cutting.

However, it is unsuitable for marking lines along the grain of the timber, and the marking gauge has been designed with this work in mind. The gauge consists of a beam, with a steel spur fixed at one end. The beam slides through a stock, along which it can be secured at any point with a thumb-screw. Thus the spur can be set, using a rule. There are a number of marking gauges with graduated measurements – a style particularly favored in North America. The gauge is usually made in beech, with brass facing strips to reduce wear.

If a great deal of marking has to be done across the grain, a variation of the marking gauge, known as the cutting gauge, could be purchased. The component parts are the same, except the spur, which is replaced with a small blade held in place by a tiny brass wedge. This blade can be honed in several ways to slice across the fibers of the timber in the same way as the knife.

This cutting gauge is most useful in marking out for stringing and banding in veneer work.

A more specialized tool is the tenon marking gauge. This has four beams, which can be set to mark up the mortice and tenon joint. One made in Germany has an inlaid graduated strip in each beam, the beams being numbered in sequence. This tool would be useful where a great deal of tenon marking out

Left, a pair of mortice gauges; the gauge on the left also has a single spur on the underside of the beam for marking out. Below, a cutting gauge, left, for marking across the grain, and a double-slide marking gauge with a millimeter scale on each beam. Right, a selection of marking gauges. 1 A heavy-duty panel marking gauge. 2–3 Four- and seven-inch/100- and 175-mm marking gauges. 4 The more complex tenon marking gauge with four beams, each numbered and having a millimeter scale. 5 A mortice and marking gauge.

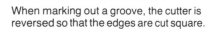

When marking out a groove, the cutter is reversed so that the edges are cut square.

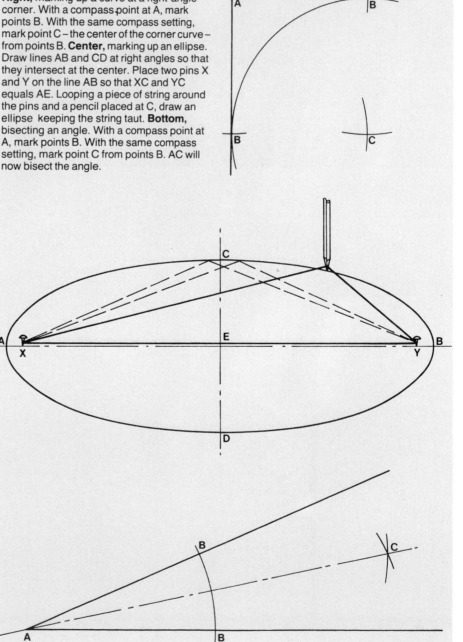

is to be done. The only alternative to this would be to use a number of separate marking gauges, numbered to ensure correct order of use.

Another variation is the twin-beamed, or double-slide, marking gauge. Both beams slide independently in a beechwood stock and have fixed, hardened, steel spurs. Easily-read scales are recessed into both beams for accurate setting.

The marking of the mortice and tenon joint, however, is usually carried out using a mortice gauge. It is similar in construction to the marking gauge, but the beam has two spurs – one fixed and one attached to the end of a sliding brass strip. The strip can be adjusted by a thumbscrew to position the spur at the required width from the fixed spur. A number of patterns also have a single spur placed on the underside of the beam, so that the tool can be used as a marking gauge. The better-quality mortice gauges are made in rosewood, with brass fittings.

The woodworker constantly gauging large timbers may decide to invest in a heavy-duty panel gauge. This is a larger version of the ordinary marking gauge.

When using the gauge, the stock must be held firmly against the timber, and the gauge pushed away from the operator. The spur of the gauge must trail – that is, the spur can be seen as the gauge is pushed along the wood.

Door-hanging seems to present a problem to many people. Often, this may be due to the difficult job of marking out the housings for the hinges. A solution to this is the butt gauge, designed with three adjustable spurs which can be set to the hinge itself. It is also useful for marking out lock and lock strike housings.

The normal pencil compass or divider will mark out small circular work, but the woodworker will have to scribe

Marking Up Techniques

Right, marking up a curve at a right-angle corner. With a compass point at A, mark points B. With the same compass setting, mark point C – the center of the corner curve – from points B. **Center,** marking up an ellipse. Draw lines AB and CD at right angles so that they intersect at the center. Place two pins X and Y on the line AB so that XC and YC equals AE. Looping a piece of string around the pins and a pencil placed at C, draw an ellipse keeping the string taut. **Bottom,** bisecting an angle. With a compass point at A, mark points B. With the same compass setting, mark point C from points B. AC will now bisect the angle.

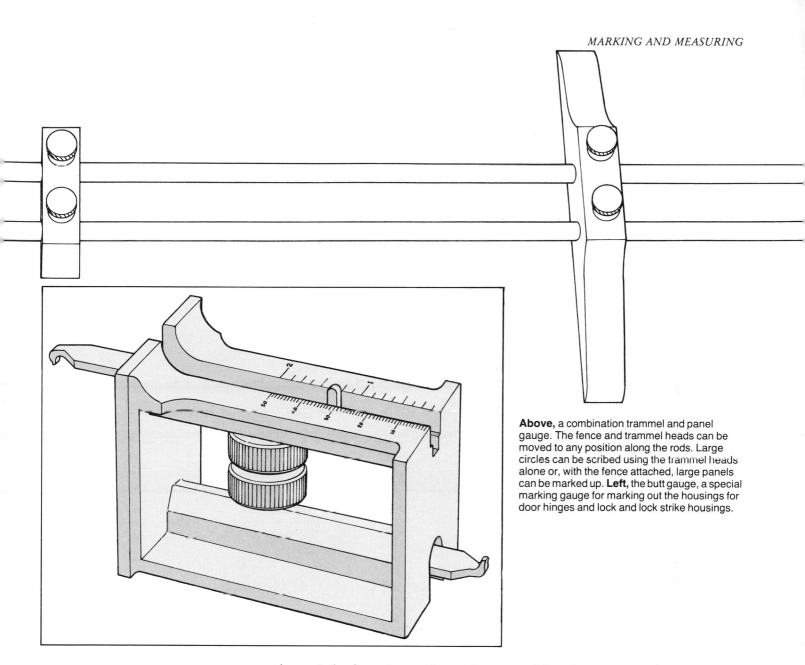

Above, a combination trammel and panel gauge. The fence and trammel heads can be moved to any position along the rods. Large circles can be scribed using the trammel heads alone or, with the fence attached, large panels can be marked up. **Left,** the butt gauge, a special marking gauge for marking out the housings for door hinges and lock and lock strike housings.

A pair of trammel heads are attached to a beam to scribe large circles.

large circles from time to time. A beam compass, which is fitted with a pair of trammel heads, will be useful. These heads slide along either a steel or wooden beam, can be fixed in any position, and are both fitted with hardened steel pins. One serves as the compass point, and the other pin can be used to scribe the circle, or a pencil can be substituted. One pattern has the pins fitted so that they can be set at an angle to mark out circles in difficult positions.

Another variation of this tool is the combined trammel and panel gauge. The trammel heads slide on two rods or beams and can be fixed in any position with knurled screws. A fence which again slides along the rods converts the tool into a panel gauge. This is a superb tool, particularly useful for marking out large panels of plywood and other man-made boards.

Often the center punch is regarded as a tool for the metalworker, but it can be used for marking out in wood and other materials. It is particularly useful for positioning screws when hingeing, and when positioning other cabinetmaker's hardware. The anvil is struck by a mallet to drive the tip of the punch into the wood. A better version of this tool, called a catapunch, avoids the use of a mallet. When the anvil is depressed, an in-built spring causes the tip of the punch to mark the wood. This is especially useful if the woodworker is holding another tool in his other hand.

All these tools, well-selected and carefully used, will ensure a good start to any project. Guard them against abuse, lightly oil those which may corrode and, at the same time, learn to appreciate the beauty of timber, from which many tools are made.

Saws

*The saw
is an essential tool in the
woodworker's kit,
and is vital when both
preparing and working
with timber.*

THE SAW IS AN ESSENTIAL TOOL in the woodworker's kit, and is vital when both preparing and working with the timber. Before choosing a saw, it is essential to understand its function, so that the correct saw is bought.

Basically, saws can be divided into three main groups – hand saws (including panel saws), back or tenon saws and special saws. Hand saws divide further into distinct types indicated by the design of the teeth – rip teeth and cross cut teeth. All saw teeth (usually called points) are set alternately to the left and right, and when pushed through wood cut a groove called a kerf.

The rip teeth are sharpened at their points and cut like a series of small chisels, to cut with the grain. This saw should never be used across the grain, or serious ripping and tearing of the timber will take place.

The cross cut teeth are sharpened on their edges, and cut two tiny grooves with the timber crumbling in between these to make a severing kerf. This saw could be used to saw with the grain, but it would be hard work.

Therefore, it is vital when buying a saw to remember its purpose, and choose accordingly.

The reputable manufacturer will have selected the best-possible steel, and will also have designed a suitable handle to fit both the hand of the user and the saw blade itself. Many woodworkers will prefer a traditional timber handle, although modern plastic is quite adequate and often pleasant in appearance. A saw must be easy to use and well-balanced – you can immediately detect a bad one. The teeth must be carefully and accurately set and sharpened – look out for the cheaper versions which have stamped-out teeth.

With the gradual disappearance of the skilled saw doctor, the woodworker must bear in mind that after some use, his saw will need sharpening, and he may have to do this himself.

Protection for all saws is necessary to prevent the teeth becoming blunt. Many woodworkers store the saw with the teeth enclosed in a groove cut in a block of wood. A number of manufacturers supply a plastic strip to protect the teeth. Although a light oiling of the blade will prevent rust, perhaps the best way to protect the saw is to keep it in a saw case. These are available in a variety of materials, some of which are waterproof. An additional precaution would be to insert a piece of rust-preventative paper in each case.

A discussion on saws would not be complete without mentioning that present-day manufacturers often coat the blades against corrosion. The Teflon coating provides protection, and some woodworkers maintain that it makes sawing easier. However, the coating must be of high quality to resist wear.

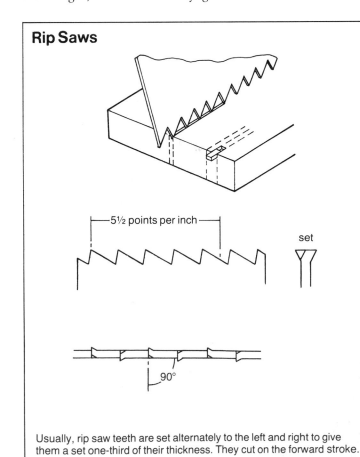

Rip Saws

5½ points per inch

set

90°

Usually, rip saw teeth are set alternately to the left and right to give them a set one-third of their thickness. They cut on the forward stroke.

Cross Cut Saws

8 points per inch

set

65° 65°

Cross cut teeth act like knives to cut two parallel grooves on both forward and backward strokes against the grain.

Log Saws

Where large timbers have to be cut, and the woodworker is unable to employ the services of the professional tree surgeon, a number of different saws can be used.

The two-man cross cut saw is superb for handling large logs. It comes in a variety of sizes and the teeth are interspersed with deep gullets to carry the sawdust. The best have a style of teeth

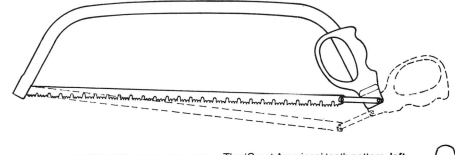

The 'Great American' tooth pattern, **left** – sets of three deep indentations, which help to throw out the sawdust – has become increasingly popular for the log saw. The modern version is illustrated **above,** and the two-man cross-cut **below.**

called lightning. The blade is curved so that the teeth can be kept in contact with the timber throughout the movement of the saw.

The one-man cross cut saw has a handle similar to that of the hand saw but with an additional round handle which can be placed anywhere along the top of the blade. Usually, it is placed at the top of the blade, near the handle, but the saw can be converted into a two-man cross cut saw by fixing the supplementary handle near the toe of the blade.

The contemporary log saw consists of a tubular steel frame with a saw blade between 24–36 inches/600–900mm long and ¾–1 inch/18–25mm in width. The blade is tensioned either by screw or lever. This tool is most useful with pegged teeth and gullets, which cut efficiently in both directions. It can tackle many rough conversion jobs which would be hard work for the normal hand saw.

A number of manufacturers fit the saw with a wrap-round handle. It is sometimes also known as the bushman's or forester's saw.

The teeth on the one-man cross-cut, **right,** are hand-filed to give maximum cut on push and pull.

Hand Saws

This group includes rip saws, cross cut saws and panel saws and, as previously described, they are distinguished by the style of their teeth. Generally, hand saws from reputable manufacturers are taper ground. This produces a blade which tapers from the handle to the toe on the back of the saw, but which retains a single thickness just above the teeth. This type of saw cannot bind in a deep cut, and the set of the saw need not be quite as great as with an unground blade. The saw is easier to use, and saves wood.

A rip saw is needed to cut a piece of timber along its length – that is, with the grain. It is the longest of the hand saws, usually 28–30 inches/700–750mm, with 3½ points per inch/25mm. Some rip saws can have smaller teeth at the toe than at the handle. There is also a half-rip version, slightly shorter in length, with an extra point per inch/25mm.

The cross cut saw for cutting across the grain is between 22–26 inches/550–650mm in length and has 5 points per inch/25mm.

The panel saw is for finer work and useful for cutting thin plywood panels. Usually, it measures 20–24 inches/500–600mm in length, with 7 points per inch/25mm. There may be variations in design between manufacturers, even in the length and number of points.

All good hand saws will have handles secured by brass screws – this is essential to take up any slack and ensure a perfectly tight union.

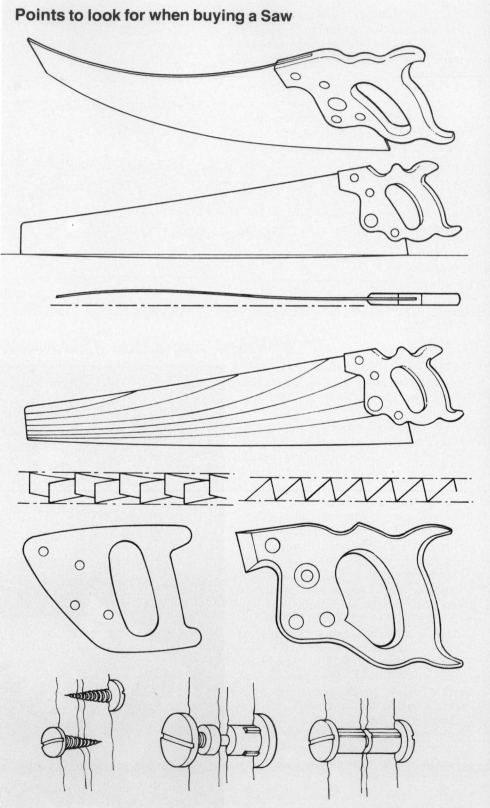

Points to look for when buying a Saw

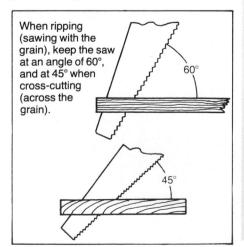

When ripping (sawing with the grain), keep the saw at an angle of 60°, and at 45° when cross-cutting (across the grain).

60°

45°

Flexibility: Although you should be able to flex the blade easily, its tension should quickly bring it back to true.

Crown: The greater the crown, the fewer the teeth that make contact with the wood. An ⅛-inch/3-mm crown on a 26-inch/650-mm blade makes sawing easier without affecting efficiency.

Straightness: Before buying, check that the blade is quite straight and lies properly in the handle.

Taper grinding: A taper-ground blade means that the teeth require less set and the blade will not bind in the kerf. The manufacturers specifications will tell you whether it has been ground in this way.

Teeth: Sharpness, of course, is essential. Make sure also that the set is uniform and that there are no burrs on the teeth.

Handle: Check that the full four-finger grip is comfortable. Higher-quality saws usually have wooden handles. These will not make the hand sweat so much, and after a while they will begin to follow the contours of the hand.

Fittings: How is the handle secured? **1** Cheap screw fixing, **2–3** better and best shoulder screw fixing. Brass fittings are best, which may be nickel-plated.

Special-Purpose Saws

For curved cutting, the frame, or bow, saw is probably the most versatile of the special saws. It has a thin, narrow blade attached by removable and renewable pins, through brass ferrules, located in the lower ends of two wooden cheeks, or arms. These are held apart by a centrally-placed stretcher rail, or beam, morticed into the arms by stub tenons and tensioned by a twisted cord worked by a toggle stick. Handles are attached to the ferrules, which can be twisted in the cheeks to position the blade in the direction the curve is to take. Blades can vary between 8–28 inches/200–700mm in length, and usually have 9 points per inch/25mm. Normally, it is used for curved cutting only. The depth of the cut is indicated by the distance between the stretcher rail and the blade.

The continental bow saw is larger, and used for all types of work. The biggest sizes have fixed wide blades, without handles, and are often used instead of the hand saw. Another version sometimes seen in Europe is the double-bladed frame saw, with an adjustable center stretcher rail, which serves to tension the blade.

A smaller saw for curves, similar in concept to the frame saw, is the coping saw. This is a modern design with a 6-inch/150-mm blade, tensioned in a steel frame by twisting its round wooden handle. The frame cuts up to 4½ inches/112mm from the timber edge. The blade has two pins which fit into swiveling spigots at each end of the

A coping saw like the one **above** has teeth which have been carefully hardened and set to give a beautifully smooth action. The smaller of the frame saws, **below,** is useful for cutting a radius. The larger one has four different blades, two for rough cutting and two for cabinet work. The swivel handles allow cutting in any direction.

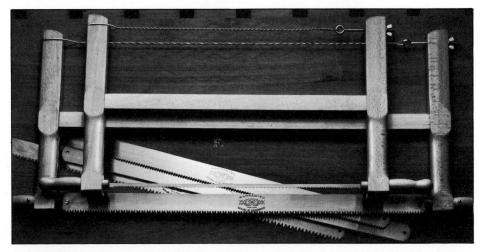

Classic handsaws, with both cross cut and rip saw teeth are contrasted with the more unusual looking bow saw. The saw case will protect hand and tenon saws when not in use.

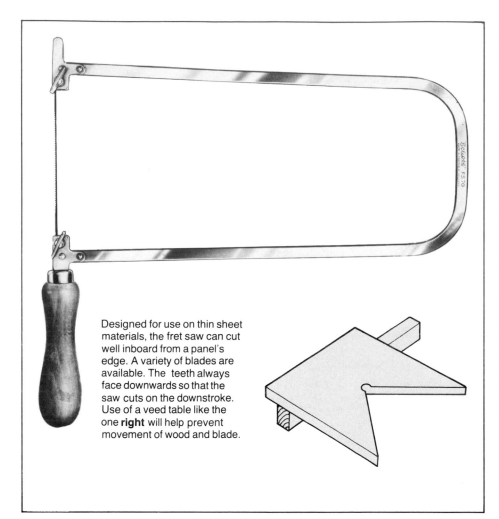

Designed for use on thin sheet materials, the fret saw can cut well inboard from a panel's edge. A variety of blades are available. The teeth always face downwards so that the saw cuts on the downstroke. Use of a veed table like the one **right** will help prevent movement of wood and blade.

frame. The ease with which the blade can be fitted makes it possible to assemble it after passing it through a previously bored hole, enabling enclosed curves to be cut within a panel. This saw is also used for the fast removal of waste wood when cutting dovetails and other joints.

Many woodworkers prefer to insert the blade so that the teeth point towards them. The saw then cuts on the back stroke, which allows straighter cutting and reduces breakages. This is certainly an advantage when cutting thin stock which rests on a veed table mounted in the vice. With the saw used vertically, the natural tendency of the teeth is to pull the wood down onto the table, again reducing the possibility of saw breakage.

A saw for small curves, slots and key-holes is the pad saw, also known as the compass, or keyhole, saw. This has a narrow tapered blade which fits into a round handle. The blade can be fixed in any position by turning a thumbscrew

The handle of the pad saw, **below,** will allow a blade to pass through wood, giving adjustable blade projection. It will accept broken hack saw blades. The saw is used for such jobs as the cutting of keyholes and letterbox holes. The general-purpose saw, **bottom,** has a blade which can be adjusted to nine different positions, and is usually Teflon-coated. It is a not really a craftsman's tool.

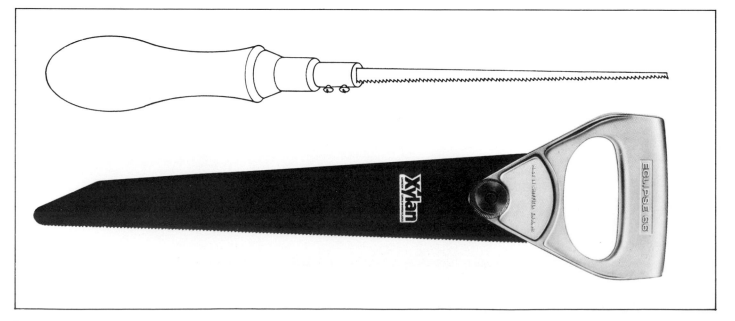

The nest of saws, **1**, has interchangeable keyhole and squared-off blades. The teeth on the upper edge of the flooring saw, **2**, allow you to cut to the edge of a board from an upright position.

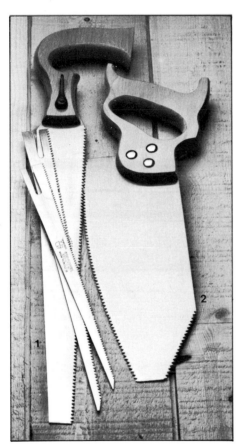

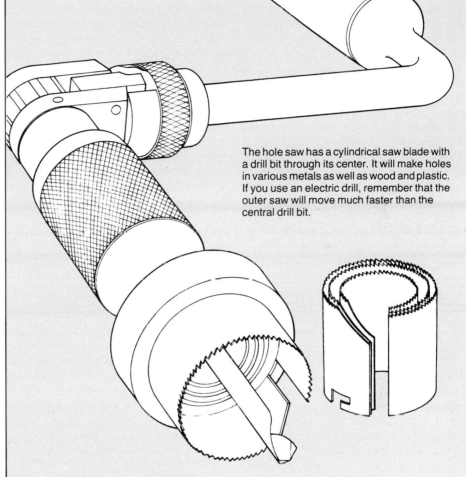

The hole saw has a cylindrical saw blade with a drill bit through its center. It will make holes in various metals as well as wood and plastic. If you use an electric drill, remember that the outer saw will move much faster than the central drill bit.

located in the handle ferrule. Blades vary between 5–15 inches/125–375mm in length, and from 8–10 points per inch/25mm. The handles of many pad saws also take a whole or broken hack-saw blade.

The pad saw blade is seen again in the nest of saws. This comprises one open-ended handle into which various tapered blades, slotted at the wide end, can be fitted. This is a useful tool for the casual woodworker.

An extremely versatile saw for fine work, particularly the piercing of thin panels and plywood, is the fret saw. It was a craze amongst schoolboys in Europe in the early part of the twentieth century. A number of versions are still available, with frames up to 20 inches/ 500mm deep, allowing cuts to be made inboard from the edge of the panel. The

very fine blade, up to 6 inches/150mm in length, is held in position by a clamp, with the tension applied through a thumbscrew. Different blades can be obtained for cutting wood, metal or plastics.

The hole saw, although not strictly in the saw family, is a very useful tool. This consists of a circular body and a drill bit, which fits into the chuck of a portable drill. Circular blades are fitted into the body and the portable drill switched on, once the drill bit has been correctly located on the material. Various sizes of blades are available, enabling holes to be cut in timber, metal and plastic.

For the craftsman working on a house, particularly when needing to gain access under floorboards for electrical or heating work, a special

flooring saw can be used. The blade is 13 inches/325mm long, with 7 points per inch/25mm. The top side of the blade is angled to allow the saw to be used in an upright position close to a corner. The other edge has teeth completely along it, and is curved to cut into the tongue of the floorboard without the need for a starting hole. Another version of this tool is the plywood saw, approximately 11 inches/275mm long. The end of the top edge is curved and toothed, thus permitting cutting into a panel without pre-boring.

A fairly new addition to the saw family is one which is fitted with a low-friction blade, capable of cutting wood, steel, plastics, rubber and many other materials. The blade is set in an enclosed handle, and can be moved into nine different positions.

Back or Tenon Saws

There are various types of saws in this group. Generally, the blades can be described as thin, short and rectangular in shape, with fine teeth finely set. A heavy strip of either brass or steel is folded over the back of the saw to strengthen it. Brass, being a heavier metal, is better than steel. The design of the saw means that it cannot penetrate the timber completely, making it suitable for cutting joints. The handle, similar to that on a hand saw, is often set higher, and is either closed or open. A smaller back saw, with an open handle, is known as a dovetail saw. It is 8 inches/200mm long, with as many as 26 points per inch/25mm. The blade is extremely thin, with finely-set teeth to give complete accuracy in the cutting of dovetail joints and other fine work.

Tenon saws are usually between 10–14 inches/250–350mm in length, with up to 20 points per inch/25mm, and are used to cut tenon joints.

Another variation of the back saw is the gent's back saw, so-called after the small tools once used by gentlemen who took up woodworking as a hobby. This has a turned chisel-like handle, with blades up to 10 inches/250mm long. The blade is thin, with small teeth, set to cut a narrow kerf. It is used best for cutting fine joints and dovetails. This saw has also been called the daney gent's saw, the gent's dovetail saw and the beading saw. Many users find this saw difficult to hold after using a saw with a normal handle. A continental manufacturer offers the gent's-style saw with an offset handle, and one with a reversible offset handle.

A variation of the gent's back saw is the blitz saw. This has interchangeable blades for use on metal, wood or plastics. A small hook as an appendage at the end of the back strip serves as an extra handle. It was not originally designed specifically for the wood-worker, but was an all-purpose saw.

Ancillary Sawing Equipment

Many sawing jobs can be carried out with the timber held in a vice, but the sawing of tenon shoulders, and many other smaller, shallower cuts in thinner timber, will be done best using a bench hook. This is a traditional piece of equipment in the United Kingdom, consisting of one large block of wood with an end block fixed on either side. Generally, it is made of beech with the end blocks doweled to prevent any possibility of the saw striking screws or nails. The bench hook is usually held in

The classic brass-backed gent's saw, **1–2**, is shown on the **facing page.** Derivations of this, with fixed offset handle, **top,** reversible offset handle, **middle,** and straight handle, **bottom,** are shown **below. 3–4, facing page,** are the blitz saw and slotting saw, useful for fine work. The back or tenon saw, **5,** cuts smaller pieces to length. **6,** the dovetail saw.

Making a Sawing Sizing Board

The sizing board illustrated here is a simple but reliable and efficient way of ensuring that a length of wood is cut in completely equal sizes. The base is made from ½ inch/12mm blockboard, with a renewable hardwood insert to protect the base. The upright section is made from beech or any other hardwood ⅝ × 1¾ inches/16 × 45mm. It has a sliding hardwood stop which can be fastened with a coach bolt and wing nut.

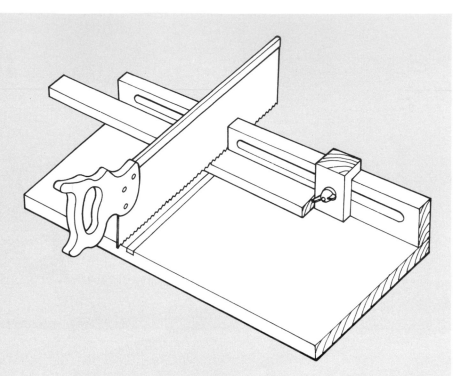

Dimensions in inches/mm

A	18/450	H	½/12
B	8/200	J	¹³/₁₆/21
C	⅝/16	K	¾/18
D	7½/187	L	2¼/58
E	⅜/10	M	¹⁵/₁₆/24
F	1/25	N	1¾/45
G	¾/19	O	⅜/10

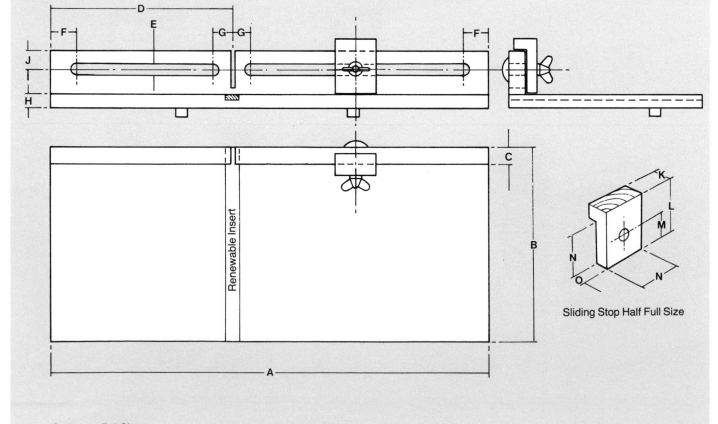

Renewable Insert

Sliding Stop Half Full Size

Scale – ¼ Full Size

Sawing Jigs

The use of two bench hooks like **1** will prevent a long piece of wood from bending too far downwards. A mitre block, **2**, will help you to cut angles of 90° and 45° quickly, and a mitre box, **3**, will also help you to keep your saw true, but it limits the width of wood for which it can be used. The metal version, **4**, incorporates clamps and ensures high precision. The mitre cutter, **5**, with the built-in saw is made to keep friction from slip bearings to a minimum. It is a worthwhile investment if a large amount of this type of work is envisaged.

The jointmaster, **6**, allows novice and expert alike to make cuts of many types and angles.

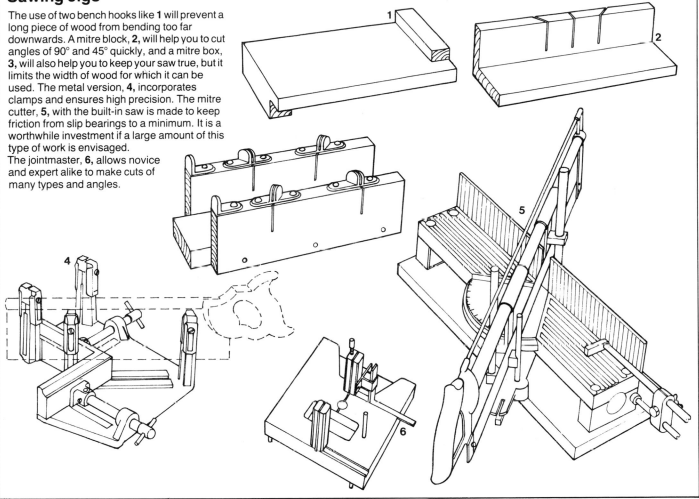

the vice, but it can be used freely hooked over the edge of the bench. Either side of the bench hook can be used.

Normally, mitres are cut with a tenon saw. The mitre can be marked out using the mitre square or the combination square. A quicker way is to use a mitre block.

This is usually made in beech, and comprises two pieces jointed at 90°, with 90° and 45° slots cut in the upstand. It can be held in the vice or screwed to the bench. The saw passes through the slot and is held in the vertical with the timber held firmly against the upstand.

A more efficient tool is the mitre box – the timber is housed between two slotted slides. The slots are sometimes reinforced with brass guides. A deluxe

metal version has saw guides which can be adjusted to the exact thickness of the blade. The largest size will take timbers up to 4×2 inches/100×50mm.

Another metal model, which not only incorporates guides for the saw but also screw clamps for the timber, ensures great accuracy in cutting. This tool is particularly valuable where careful cutting of picture frames is needed. An extended model has sawing positions for both 45° and 90°. Made in grey iron with accurately machined faces, this tool can also be used for joint work.

A greatly superior mitre cutter is one which has its own built-in saw assembled on easy-moving bearings. It can be quickly adjusted to cut at 90°, 60° and 45°, and incorporates an adjustable

stop for measured cutting. Unfortunately, it does not have a timber-clamping device.

An increasingly popular tool is the jointmaster. This tool was designed to cut lap joints, mitre joints, tenon joints and many other cuts at degrees varying between 10° and 170° using an 12-inch/300-mm back saw. It accommodates quite large timber and can be used with success by even the most inexperienced woodworker. The timber is held in place by wedges and pins.

The dedicated woodworker will devise many different ways of holding for sawing. A device used by the author for many years for repetitive sawing is the sizing board. This is made in blockboard or multi-ply wood.

CHAPTER FIVE

Planes

*Planes are used to
prepare the surface of timber, for
cutting joints and cleaning
up finished work.*

P

LANES ARE USED TO PREPARE the surface of timber before joints and other constructions are marked out. They are needed also for cleaning up finished work as well as for cutting joints.

As described in Chapter One, these needs were met in the eighteenth and nineteenth centuries by an infinite variety and shape of plane, all constructed of wood. Luckily, this position changed with the introduction of metal planes, many of which were multipurpose, so the woodworker needed fewer planes to carry out the same jobs. Improved wooden-bodied planes are still available, much to the delight of those woodworkers who prefer them.

Another group of edge tools, similar in action to bench planes, are spokeshaves and drawknives. Scrapers are also included in this section, because they are finishing tools (although with a different action to the smoothing plane, which is the most commonly-used finishing tool).

Planes can be divided into two groups – bench planes, used mainly for truing and sizing timber, and special planes, designed to make specific cuts and to shape timber for joints. The latter group are sub-divided further into four categories: (a) block planes, (b) rebate, plough and multi-purpose planes, (c) shoulder and bull nose planes, and (d) planes for special trades.

Any discussion on planes must begin with a description of the cutting action. The essential part of any plane is a cutter or blade made of a material which keeps a sharp edge. The cutter must be held squarely in the body and be adjustable, so that its edge always protrudes a hairline's thickness below the sole of the plane. The angle at which the cutter is ground and sharpened, and at which it is held in the plane body, is extremely important, if the best results are to be obtained. At the same time, care must be taken to control the quality of cut, so that the surface of the wood is left smooth and flat, and the shavings leave the plane without clogging it.

Bench planes are fitted with a cutting iron, which has a cap, or back, iron attached on the flat side. This breaks and rolls the shaving, which is then ejected from the plane escapement. The closer the cap iron is to the cutting edge, the sooner the breaking of the shaving begins.

The shaving is prevented from tearing ahead of the cutter by the closeness of the forward edge of the mouth.

The metal plane offers mouth adjustment through the re-positioning of the frog, so that very fine cutting can be achieved even on the most cross-grained of timbers.

An extremely wide mouth will give a result similar to that of a plane working without a cap iron.

However, if the cutter is adjusted too far, a thicker shaving will only choke the mouth and stop the plane working.

Planes designed to cut end grain fall into two of the special categories – block planes and shoulder planes. In these planes, the cutters are set at a much lower angle with the ground bevel of the cutter on top, to give what is almost a slicing action, and the bevel itself rolls over the shaving. These planes have narrow mouths or mouth adjustments to ensure the quality of the cut.

The adjustment of cutters was a matter of trial and error in the old wooden planes, but in most metal planes and the contemporary wooden planes, full and accurate adjustment fore, aft and laterally is possible.

Some of the present-day wooden planes are still without adjusting screws, the cutter being held in place with a simple wedge, and adjustment made by tapping the cutter with a hammer.

Most manufacturers supply metal bench planes with corrugated soles, which reduce suction between the face of the board and the sole of the plane. The corrugations allow just enough air to break the vacuum when the plane is in use.

Adjusting a Metal Bench Plane

To adjust the frog, slacken screws A, make adjustment by turning screw B, retighten screws A. The mouth is set wide for coarse work, narrow for finishing and fine work.

To set the cutting edge parallel with the sole, adjust the lateral lever.

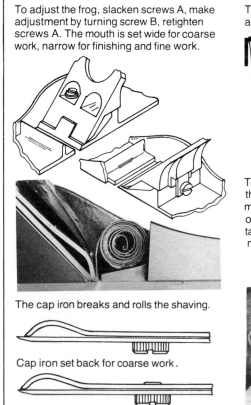

The cap iron breaks and rolls the shaving.

Cap iron set back for coarse work.

Cap iron set almost level for finishing.

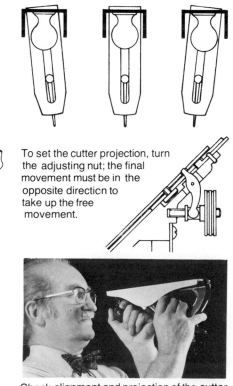

To set the cutter projection, turn the adjusting nut; the final movement must be in the opposite direction to take up the free movement.

Check alignment and projection of the cutter.

Cross-Section of a Metal Bench Plane

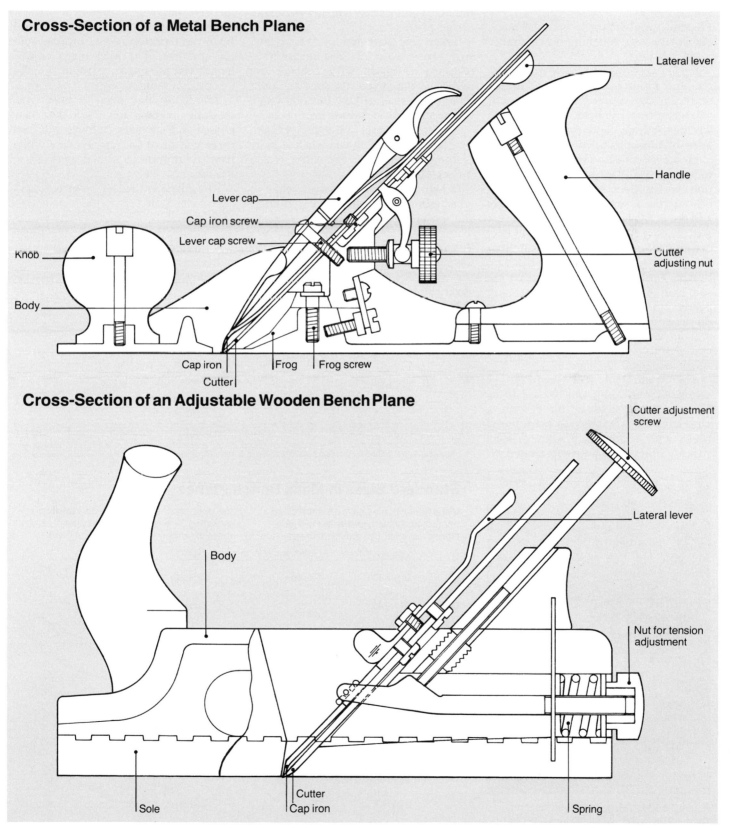

Lateral lever

Handle

Lever cap

Cap iron screw

Lever cap screw

Knob

Cutter adjusting nut

Body

Cap iron

Frog

Frog screw

Cutter

Cross-Section of an Adjustable Wooden Bench Plane

Cutter adjustment screw

Lateral lever

Body

Nut for tension adjustment

Sole

Cutter

Cap iron

Spring

Bench Planes

The first popular plane was the scrub plane, used for roughing down timber from the saw.

Earlier versions had single irons. The modern scrub plane has a red beech body and white beech, or hornbeam, sole. The cutter is held in place with a wooden wedge, and the plane has a horn-like front handle.

However, most woodworkers prefer to use the jack plane for timber preparation. Its length of 14 inches/350mm has become the norm for schools, colleges and the home craftsman. The metal jack is made of grey iron, with brass and steel components. The cutter is of tungsten vanadium steel, and the handles are beech. These planes are ground on the sole to extremely fine limits and produce a beautifully polished surface on the timber. The cutter of the 05 plane is 2 inches/50mm wide but if a wider cutter is necessary, use the 05½, which is 2⅜ inches/60mm wide.

The traditional jack plane is 17 inches/425mm long, and has a 2-inch/50-mm cutter held in place with a wooden wedge. It is usually made in red beech with white beech soles. A steel striking button has been substituted for the boxwood button which was used in earlier planes.

The trying plane is a long jack plane, and varies from 22–25⅜ inches/550–640mm in length, with a blade 2⅜ inches/60mm wide. It has a closed handle and steel striking button, with a red beech body and white beech sole.

The fore plane, 18 inches/450mm long, should be used when longer planing is required. The cutter is 2⅜ inches/60mm wide.

When jointing long boards, edge to edge (sometimes called rub jointing), the jointer bench plane must be used. The 22-inch/550-mm long plane has a 2⅜-inch/60-mm wide cutter, and the 24-inch/600-mm long plane has a 2⅝-inch/66-mm wide cutter. The latter is the longest metal plane on the market. Both these planes have reinforced soles for stability. The long plane ensures lengthwise accuracy in planing, which is essential in jointing work.

The smoothing plane is used when cleaning up completed work. The plane is available in three body lengths and three widths of cutters, so that anyone, from schoolboy to skilled worker, can use it.

The plane is sold in great numbers,

Right, a selection of wooden planes with wooden wedges for cutter adjustment. **1–2** 17- and 9½-inch/425- and 237-mm jack planes. **3** A trying plane. **4** A smoothing plane. **5** A block plane, used for cutting end grain.

Some bench planes are made with corrugated bases – these prevent suction forming between the sole and the material and assist free movement when working with resinous timber.

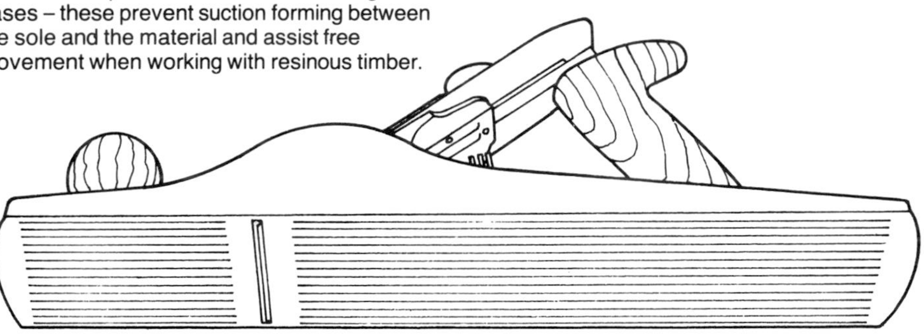

Planing long edges, guide with the fingers touching the wood, the thumb pressing down

Standard Sizes in Metal Bench Planes

Metal bench planes have a standardized numbering system, so that all No 5 jack planes, for example, are the same size. In the past, other sizes have been available, including 01 and 02 smoothing planes, but there is no longer any demand for them.

No	Length in/mm	Cutter width in/mm	Type
03	9½/240	1¾/45	Smooth
04	9¾/245	2/50	Smooth 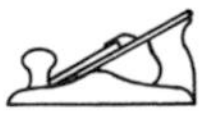
04½	10¼/260	2⅜/60	Smooth 
05	14/355	2/50	Jack 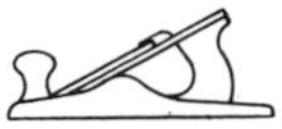
05½	15/380	2⅜/60	Jack 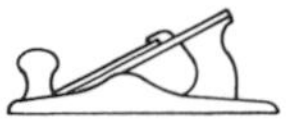
06	18/455	2⅜/60	Fore 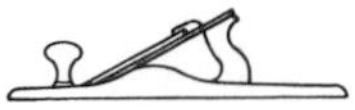
07	22/560	2⅜/60	Jointer 
08	24/610	2⅝/70	Jointer 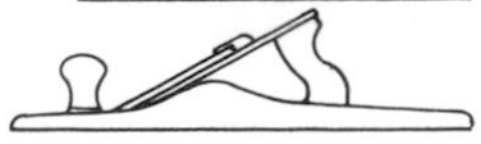

and is probably the most used and abused of all the bench planes. There are many cheap versions, some of which are manufactured without any frog adjustment.

It is made of red beech or cast iron. The wooden plane has a horn-shaped handle at the forward end, a double iron and a wooden wedge. A block placed immediately behind the cutting unit provides comfortable placement for the hand. These planes vary from 6½–9½ inches/162–237mm in length, and are about 2 inches/50mm wide. Metal smoothing planes are from 9–10¼ inches/225–256mm long and 1¾–2⅜ inches/43–60mm wide.

The fully adjustable, traditional wooden planes manufactured in Germany under the name Primus are the best in the world. They have all the refinements of the metal planes with the warmth and comfortable feel of wood. They are made of beech, with soles of *Lignum vitae*, and are planes to last a lifetime. *Lignum vitae* is a tropical, hard-wearing wood, full of a natural oil which helps the plane run smoothly.

In the future, owing to lack of supplies of the traditional woods, wooden plane manufacturers will have to use new timbers, which must be hard-wearing and even-grained, and chosen with great care. Recently, a number of planes have been made of *Goncalo alves*. This is a beautiful tropical hardwood, with some of the qualities of *Lignum vitae*, including an oil secretion, but it is a much more attractive timber.

A slight variation on the smoothing plane is the reform plane, which has an adjustable front shoe for refined setting of the mouth. This smoother is regarded as the ultimate in wooden planes, and is essential for the craftsman who wishes to produce the best-possible work.

Other bench planes are built for particular tasks. The circular, or compass, plane, is designed for use on concave and convex surfaces. The flexible steel sole can be set to any curve using the center screw adjuster. The cutting unit adjustment is the same as that on other bench planes, but there is no mouth adjustment. One plane has a handle and another is shaped to fit the hands.

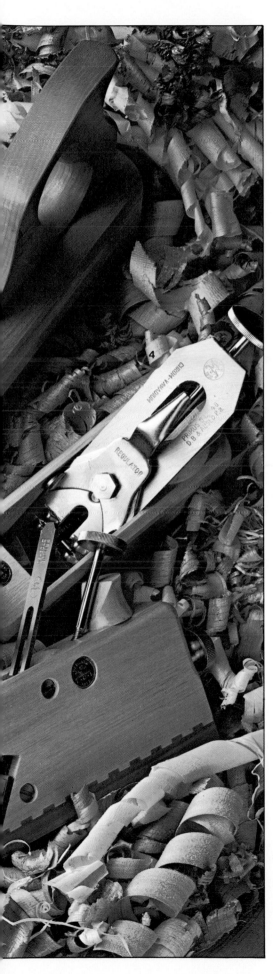

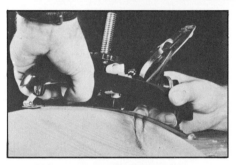

Opposite: 1 The replacement-blade plane is a recent development; the model is supplied with six throw-away blades and has a Teflon-coated sole. **2–5** A selection of wooden planes with spring-loaded cutter adjustment and *Lignum vitae* soles. **2** 9-inch/225-mm reform-type smoothing plane with adjustable front shoe. **3** 25⅝-inch/641-mm trying plane. **4** 9½-inch/237-mm jack plane. **5** 11-inch/275-mm rebate or rabbet plane with an adjustable front shoe. **Above,** a fine-quality 8⅞-inch/221-mm reform-type smoothing plane with a pearwood body, *Lignum vitae* sole, spring-loaded cutter adjustment, and an adjustable front shoe. **Left,** the compass plane has a thin, flexible steel sole for planing concave and convex shapes. The sole is located at the heel and toe of the body and a screw adjustment raises or lowers the center.

Some metal planes designed for making specific cuts or for shaping. **1** A rebate or rabbet plane. **2** A plough plane; the fence can be fitted either side of the body and interchangeable cutters will cut grooves between ⅛ and ½ inch/ 3 and 12mm wide. **3** A shoulder plane with machined sides for trimming shoulders and rebates. **4** A fully adjustable block plane. **5** A spokeshave for shaping curved surfaces; the cutter is fully adjustable.

Block Planes

A tool closely related to the bench plane is the block plane, used for cutting end grain. These are small planes, from 6–8 inches/150–200mm in length, and are designed to be held in one hand. The cutting iron can be set as low as 12°, and is used bevel-side uppermost.

One plane has a cutter which is adjustable fore, aft and laterally, and another model has two mouths, giving normal and bull nose cutter positions. Others are available with or without cutter adjustment.

A traditional wooden block plane is coffin-shaped, and made of red beech. The double cutter is held by a wooden wedge.

When using a block plane, **right**, guide the sole with the forward hand.

Cross-section of a fully adjustable block plane. The cutter, bevel-side uppermost, can be adjusted laterally and for depth; there is no cap iron. The mouth opening is also adjustable.

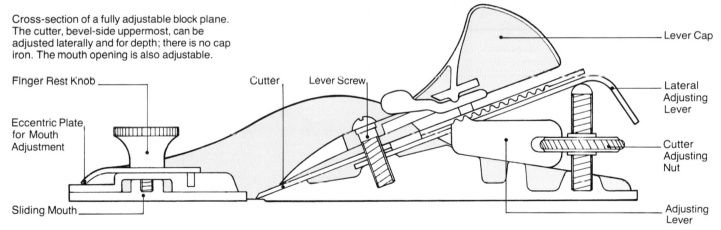

Finger Rest Knob

Eccentric Plate for Mouth Adjustment

Sliding Mouth

Cutter

Lever Screw

Lever Cap

Lateral Adjusting Lever

Cutter Adjusting Nut

Adjusting Lever

A plane which has appeared on the market only recently is the edge trimming block plane, constructed in either cast iron or manganese bronze, with a very high-quality finish. The plane has a V-shaped underbody one side, which houses the cutter, and the other side acts as a fixed fence. The cutter is housed in a similar way to that of the side rebate plane, but is angled slightly forward to give an ideal shearing or slicing cut. As its name implies, the plane will trim edges up to ⅞ inch/21mm thick both with and across the grain, exactly square. Its superb cutting action also means it can be used on man-made boards like plywood and multi-board.

The edge trimming block plane will square wood surfaces up to ⅞ inch/21mm wide with or across the grain.

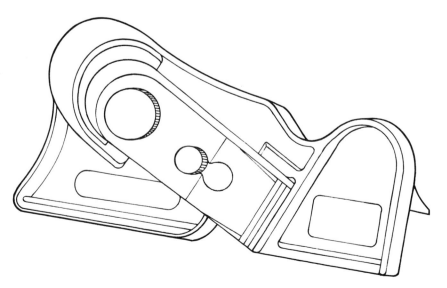

Rebate and Fillister, Plough and Multi-purpose Planes

The bench rebate, or rabbet, plane is a bench jack plane with a blade modified for cutting rebates. With metal bench rebate planes, the blade extends the full width of the sole, whereas wooden bench rebate planes have an angled blade breaking through the body on one side.

Purpose-built rebate, rabbet, or fillister planes are also available with metal or wooden bodies. The metal rebate plane is particularly useful for the craftsman working constantly on rebates and fillisters up to 1½ inches/ 37mm wide. It is made of cast iron, with two fence arms and two positions for the cutter – the normal position, where it is fully adjustable, and the bull nose position, where it has to be adjusted by hand before the lever cap is tightened.

The plane is fitted with a steel spur which severs the top fibers before the blade when cutting across the grain. The fence can be set across the cutter to limit its width, or it can be transferred to the left-hand side. Sometimes, this plane is called a moving fillister. With the fence removed, it can be used as a simple

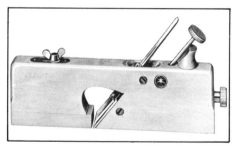

A wooden rebate plane with adjustable cutter.

square plane. One version has a single fence arm and a lever-type cutter adjuster.

Wooden rebate planes are made of beech, with a wedged cutter and steel striking knob, which is the only departure from the traditional style. Primus make a particularly good rebate plane, with a fully adjustable cutter. To improve cutting, the plane is fitted with an adjustable front shoe which allows the mouth to be set accurately. It is made of white beech, with a sole of *Lignum vitae*. Improved handling is

1 A wooden hand router, used for cutting grooves and dados. **2** A nineteenth-century style plough plane with an adjustable fence; the plough plane is designed primarily to cut grooves, but will also cut rebates. **3** A tongue and groove plane set; the tonguing plane cuts a tongue which fits tightly in the matched flat-bottom groove cut by the groove plane. **4** A traditional wooden rebate plane.

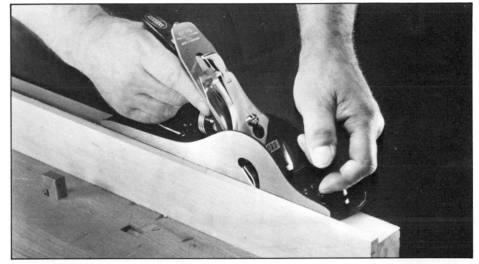

Left, the bench rebate plane is similar in design to other bench planes except that the blade extends across the full width of the sole. There is no fence guide, so a piece of wood has to be attached to the workpiece to act as a guide; this can be removed after the rebate is begun.
Below, the rebate or fillister plane is purpose-built for cutting rebates: it has an adjustable fence, a depth gauge, and a forward position for the cutter for planing up to a stopped rebate.

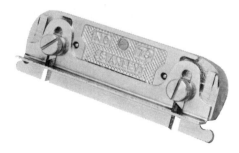

obtained when a curved block is set immediately behind the cutter.

The stop rebate plane, sometimes called the chisel plane, is a further refinement from the same manufacturer. It is cut off at the mouth so that the cutter extends in front of the body, making it possible to work right into the corner of a rebate or stopped chamfer.

The plane is made of white beech, and the cutter is held by a wooden wedge.

Whenever a housing or groove has been cut slightly too narrow, the unschooled worker will resort to using a chisel to cut the sides until the panel fits. This can be disastrous, because great care must be taken to avoid over-cutting or uneven cutting.

The side rebate plane is designed to eliminate this hazard. The plane illustrated here has two cutters which enable the plane to be worked in either direction, but they have no adjustment screws and have to be set by eye. The plane can be converted to a side chisel or side bull nose plane by removing the nose pieces. The depth gauge controls the cut.

The plough plane is an essential tool for every serious woodworker. It is so simple to set up, that it is often faster to use than it is to assemble the supposedly quicker router or routing machine.

The traditional plough plane is made of wood, and built in the style of those used in the nineteenth century. Usually, it is supplied with six cutters which are held in place with a beech wedge in a beech body. It is fitted with a steel depth gauge. The fence is adjustable along two fence arms made of pearwood, and is firmly fixed at any distance up to 5 inches/125mm from the blade.

A new plough plane was designed recently by Record Ridgway, which considers modern production methods and the needs of the woodworker. Many parts are interchangeable with other planes in the range. A unique depth-setting device and full adjustment for the ten cutters is provided. The tool is easily used by both right- and left-handed workers. It offers a wide choice of groove and rebate cuts.

The combination plane is more advanced, and not only offers all the features of the plough plane but has tonguing and beading cutters as well, making eighteen cutters in all. Spurs are provided on the body and sliding section to sever the fibers before the blade when making cuts across the grain. The plane has a bead stop to cut beads on tongued stock.

Tongue and groove sets are sold in pairs to make close-fitting panels. The tongue plane cuts a tongue which fits into the groove cut by the groove plane. These planes are made of red beech, with steel cutting irons.

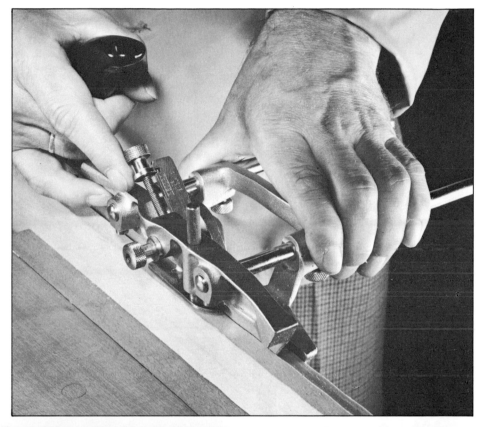

Right, cutting a rebate with a plough plane. The cutter fitted is wider than the rebate required and the width of the rebate is set by the distance between the fence and the outside edge of the cutter. A more sophisticated version of the plough plane is the combination plane, **below,** designed to perform a wide variety of planing operations. **Below top right,** cutting edge beading. **Below bottom right,** cutting square and round moldings.

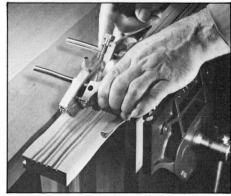

The multi-plane is made only by Record Ridgway. It is an expensive but very versatile plane, which must have a place in all reference books, if not in the average workshop. It carries out all the work covered by the plough and combination planes previously mentioned, and a large number of additional cutters and bases are available to extend its use further. This plane is capable of ploughing, rebating, housing, tonguing, fillistering, beading, sash molding and slitting. Spurs help when cutting across the grain, and these can be housed when not in use. Long and short fence arms are supplied, which can be fitted in two positions in the fence.

There is a fully adjustable depth gauge in the sliding section. Three pronged spurs which cut across the grain are housed in the body and sliding section. In addition, the plane has a bead stop, cam steady and a slitting cutter, which has an adjustable stop.

Special bases are available to cut hollows and rounds of various sizes. Stair nosings can also be cut. These bases slide over the fence arms, replacing the sliding section.

The multi-plane will perform a wide range of cuts, **above.** Special cutters, **left,** are available in addition to the standard 24, and the special bases illustrated **below** will cut hollows, rounds and stair nosings.

The Multi-Plane

The multi-plane is made of nickel-plated steel and rosewood. Its components include an adjustable fence and depth gauge, two sets of fence arms (long and short), a beading stop, slitting cutter, sliding section, depth gauge, cam steady, and spurs for cross grain work. Except for the two smallest, all the cutters are fully adjustable

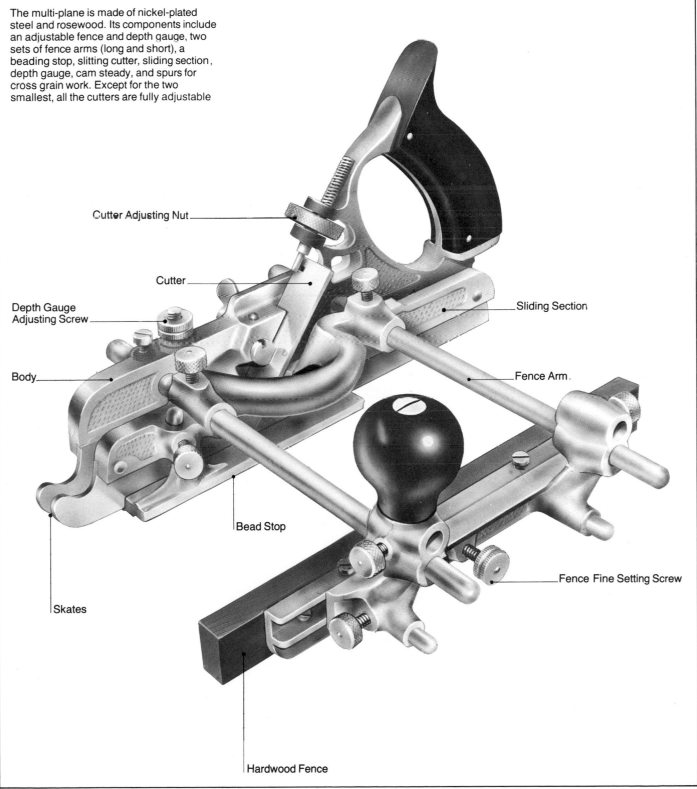

Cutter Adjusting Nut

Cutter

Depth Gauge Adjusting Screw

Sliding Section

Body

Fence Arm

Bead Stop

Fence Fine Setting Screw

Skates

Hardwood Fence

Shoulder and Bull Nose Planes

The best tool for finishing end grain, or tiny rebates, is the shoulder plane, and any craftsman working secret dovetails or lapped dovetails will need one. The thin cutter is set at a low angle, and is usually fully adjustable. The planes are made of cast iron and are finely and accurately ground on base and sides. The mouth can be adjusted by moving the body section forward.

A variation of the shoulder plane is the 3-in-1, so-called because it converts into a bull nose and a chisel plane. The front extension is removed, and an additional nose is fitted to convert the plane into a bull nose. Without either of the extensions, it becomes a chisel plane.

As in the shoulder plane, the cutter is set at a low angle, with bevel uppermost, and is fully adjustable.

The bull nose plane is used when planing up to stopped rebates, and in other work where extreme accuracy is required. It can be adjusted in the same way as the shoulder plane.

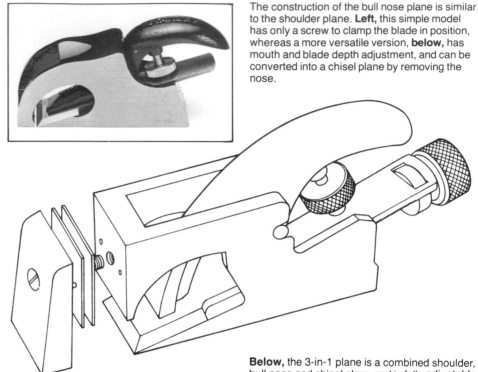

The construction of the bull nose plane is similar to the shoulder plane. **Left,** this simple model has only a screw to clamp the blade in position, whereas a more versatile version, **below,** has mouth and blade depth adjustment, and can be converted into a chisel plane by removing the nose.

Below, the 3-in-1 plane is a combined shoulder, bull nose and chisel plane and is fully adjustable.

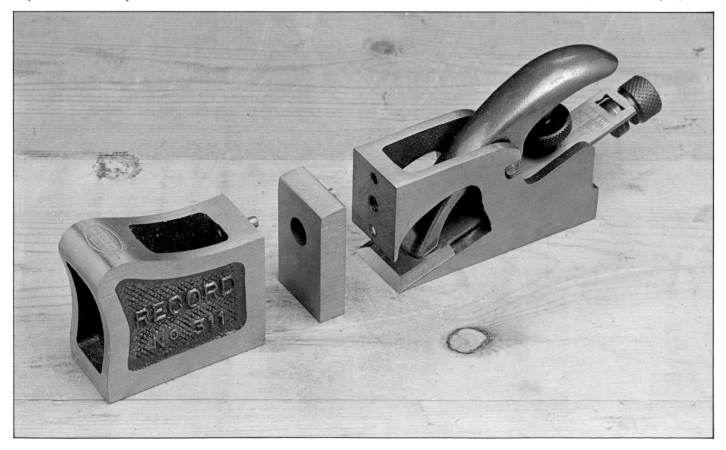

Routers

The router is one of the most useful of the special planes. It is designed to cut stopped, through and curved grooves. It is an ideal tool for cleaning ground work in low-relief carving and, at the same time, the perfecting of stopped and through housings, or dados, can be done as well.

The plane is controlled by two handles. The adjustable cutter is used in two positions – one for general use, and the other for bull nose planing. It is fitted with a fence which slides into grooves in the base to give both right- and left-hand positioning, and which has been designed to be used on both straight and curved edges. The depth stop can be used as is, running freely in its housing, with the depth controlled by a shoe in the top of the depth stop spindle or, with shoes attached, is reversed in its housing and converts the plane to a closed mouth. Narrow timbers can be grooved easily without the cutter digging in as the plane tips forward.

There are three cutters – a V-shaped smoothing cutter which has a spear-point to give a perfect slicing action, and two chisel cutters, ¼ inch/6mm and ½ inch/12mm wide.

It is not generally recognized that this is the only tool which cuts both curved and stopped grooves.

A much smaller version of the router has a ¼-inch/6-mm wide blade which can be turned in its housing to give normal or close-up work positions. It is ideal for small, sensitive work.

An 'old woman's tooth' was the name by which the original wooden routers were known. They were often rather crude blocks of wood with centrally-placed cutters. There was never any attention paid to either hand comfort or good looks.

The modern wooden version, however, is a beautifully-made tool of white beech. It has three cutters, like the metal router, but has no fence or screw adjustment to the cutter. Nevertheless, it is most useful for dado and low-relief carving work, particularly since its cutter can be turned in any direction as the job proceeds.

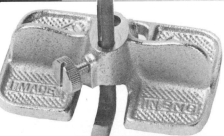

Above, cutting a housing with a router. Housings wider than the blade fitted can be cut by making more than one pass. **Left**, a miniature router is used for fine, delicate work. The two cutter housings of this model allow for stopped and through housings to be cut. **Below**, an old, hand-made router made from mahogany and brass.

Planes for Special Trades

In any major discussion on planes, those used by craftsmen making musical instruments must be included. This is usually extremely delicate work, and the removal of the thinnest shaving in a very small area is often all that is needed.

In the past, violin-makers made small planes to suit their particular needs, and had the blades fashioned by the local blacksmith. Many of the recognized manufacturers began to make these small planes in beech and boxwood.

The modern solution to this problem is the finger plane. Usually, this is made of brass, to a very high specification. The cutter is set manually and held in place by either a rosewood or ebony wedge and a steel crosspin. The soles are either flat or convex, and the body shape varies according to the manufacturer. One range is cast in the shape of whales. However, the bodies are always 1–3 inches/25–75mm long, and the blades are up to 7/8 inch/21mm wide. One brand has a toothing blade for veneer work.

These planes are designed to be held in one hand, with the forefinger resting on the front of the tool and the other fingers wrapped around it.

Not only are these beautifully-made planes essential for the violin-maker, but they are also useful for the skilled craftsman.

A small number of miniature planes are available, made of exotic timbers. They do not conform to the strict specifications demanded by the present-day violin-maker, but they are well-made and certainly will meet the needs of the model-maker.

Another one-handed tool is the palm plane. This plane has a wooden handle fitted at the end of a steel rod, and looks particularly difficult to hold. However, if the body is grasped between the thumb and fingers, the handle fits perfectly into the palm of the hand. This tool is easily controlled and the sensitive work it produces will be appreciated by wood craftsmen of all kinds, but particularly the model- and instrument-maker.

An earlier version of the palm plane looked like a miniature block plane, with an extended rear handle. It had curved and flat soles. A flat-soled plane similar to this is available now, with or without a handle.

Opposite, finger planes, originally made for violin-makers, will produce ultra-fine work in small areas; the bodies of these are solid brass. **Above,** these miniature trimming planes are made from *Goncalo alves*, an exotic tropical hardwood, and are used by cabinet- and model-makers. **Left,** a jointer-type plane, above, and a combination spokeshave and small smoothing plane made from *Goncalo alves*.

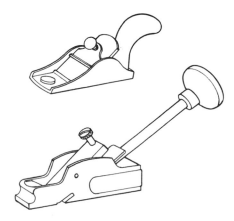

Palm planes are particularly useful for making models and instruments. The **top** one, similar to a small block plane, is available with or without a small handle, while the **lower** plane has a steel rod and wooden handle which fits into the palm of the hand.

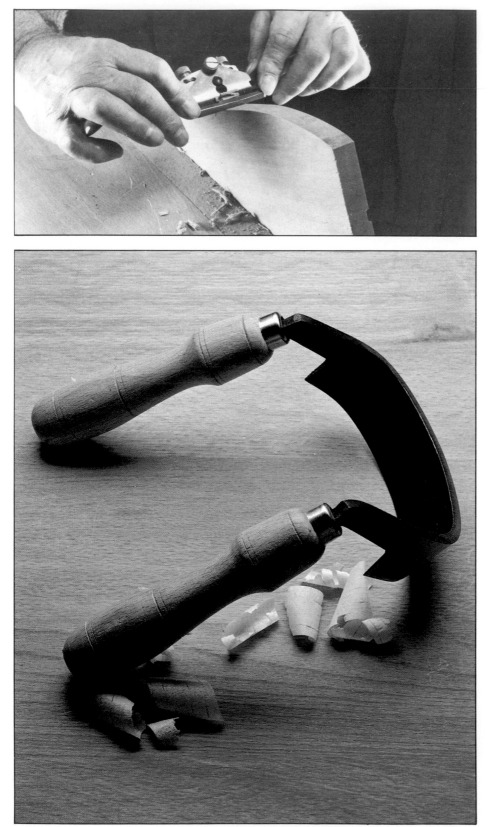

Spokeshaves

There are many jobs which cannot be tackled with a plane because of size, accessibility or peculiarity of design, but they can be done with a spokeshave.

This is really a small plane, with handles set on both sides for control. The cutter is set in the same way as with the plane, but it has no cap iron. There are flat-faced spokeshaves for convex surfaces and round-faced ones for curves.

Usually, the metal spokeshave body is constructed of malleable iron to reduce the chance of breakage.

Combination spokeshaves have one flat and one round sole and do not cost much more than a single spokeshave, but they are not as easy to handle.

A useful addition to the spokeshave range is one with an adjustable mouth.

Before buying, it is advisable to hold the spokeshave in order to judge its possible performance – the handles should fit comfortably into the palms with the forefinger at the front, alongside the mouth. Good spokeshaving is largely a matter of correct wristwork, with the front of the tool pressed down with the fingers.

Drawknives

The drawknife was the forerunner of the spokeshave but, unlike this tool, it is drawn towards the worker, has no sole and therefore requires greater skill in use. This was and still is the tool of the wheelwright, but it is suitable for a variety of other uses.

Drawknives can be used by the wood-carver and sculptor for roughing down, and the cabinetmaker and the

The drawknife, **above,** is one of the simplest shaping tools and can be used on large pieces of timber. A variation is the inshave, **left,** designed for hollowing out.
A metal spokeshave,**top left,** with blade adjustment screws. Always work with the grain and keep edges square.

chairmaker in shaping seats. Fine work is best carried out with the bevel facing downwards, and roughing down is best with the bevel uppermost.

English, Continental and American drawknives are all slightly varied in design, but the best examples have blades which taper from the back to the edge, with tangs passing through the handles and riveted over. Handles should be offset outwards for better control, and to allow the hands to clear the work safely.

A variation of the drawknife is the inshave, which has a tightly-curved blade and is designed for hollowing out.

It is particularly useful for making chair seats and other in-curving work.

Shooting Boards
A useful addition to the plane kit is the shooting board, consisting of two boards, usually beech, fixed together to form a rebate.

Two types are available – one for cutting at 90°, and the other for angles of 45°. The wood which is to be planed is laid across the top board and rests against the stop. The end grain is smoothed with either a jack or jointing plane, which is held on its side against the board.

Many woodworkers like to make their own shooting boards, and details for 90° and 45° boards are illustrated on the next page. The measurements can be adjusted to suit individual requirements.

A useful mitre board for planing a mitre along the edge of a board rather than across the end is known as a donkey's ear shooting board. The timber rests on a table set at 45°, the board itself being held in a vice.

When an end has to be planed square, use the 90° shooting board. A waste block may be placed between the wood and the stop to prevent the back edge splitting.

Making a Shooting Board

A shooting board is easily constructed from two lengths of hardwood. A chamfer is cut along the bottom edge of the upper board to clear shavings. A wedge-shaped stop is fitted in a housing in the upper board, but ensure that the edge against which the work rests is square with the board. Candlewax rubbed onto the lower board will let the plane slide smoothly.

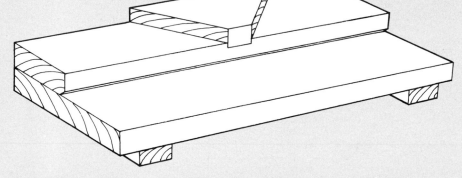

Mitres can be planed by either modifying the standard board by placing a 45° wedge against the stop or, better, by making a purpose-built mitre shooting board. The stop is placed in the middle of the upper board with its edges cut at 45°; this enables both sides of the stop to be used.

If the mitre to be cut is across the thickness of the work, as with the sides of a box, a shooting board known as a donkey's ear can be made. The upper board slopes at an angle of 45° and a rectangular stop is placed in the middle. The board is held in a vice.

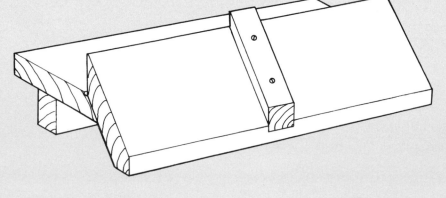

Scrapers

The cabinet scraper is used to produce a very smooth finish to wood. It is either a straight-edged sheet of steel, or one that has been shaped to fit a particular molding. This tool is held in both hands and flexed with the thumbs at the center. The edge has a minute raised burr which shaves the wood when the scraper is angled forward and pushed away from the worker. This can be achieved only with a perfectly sharpened tool, correctly angled to allow the hook to enagage the wood.

To sharpen it, the scraper is laid flat on a bench and the back of a gouge is drawn across its edge to produce a burr. The scraper is turned so it stands upright on the bench, and the gouge is worked over the edge to turn the burr outwards to the correct angle.

This is the craftsman's answer to that almost impossible piece of wood which

will not plane to a smooth, flat finish, because of its interlocking grain. A variety of straight-bladed scrapers will answer most needs, but the craftsman can always re-shape an existing one to suit a job, if necessary.

A refinement of the cabinet scraper is a double-handed scraper. This tool looks rather like a large spokeshave, and has a wide, flat sole, with a double-ended scraper blade held in place by a plate. The curvature of the blade is adjusted with a centrally-placed screw.

The blade is sharpened in the same way as the ordinary scraper. The tool is pushed away from the woodworker when it is being used, and should produce very fine shavings.

Double-handed scrapers are ideal for cleaning up veneered timbers or furniture which has been damaged through use. As the veneers are so thin, they would not withstand the use of a plane.

Adzes

It is difficult to establish whether or not the adz was the earliest wood-preparing tool. However, it has been in use for centuries, and is still used today.

The present-day adz has a head of steel and a tapered rectangular poll in which the hickory handle rests. The handle is curved so the adz edge is presented to the wood with complete control. Although the adz can remove large pieces of wood at a time, it can also finely shave timber when used correctly.

Traditionally, adzes have been used to reduce timber to size, and to trim and finish it. The results of their work can be seen on beams in houses and churches in many parts of the world.

The modern two-handed adz weighs about 5 lb/2kg. The sculptor's adz has a shorter handle, and is lighter in weight with a blade about 2 inches/50mm wide.

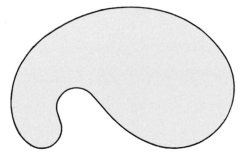

The double-handed scraper **above** is used in the same way as the cabinet scraper but is more comfortable to hold, relieving the strain on the thumbs. As the blade is held at an angle, altering its curvature controls the shaving depth. **Left**, curved scrapers are used for shaped work. The cutting edge extends all the way round, so any section can be used.

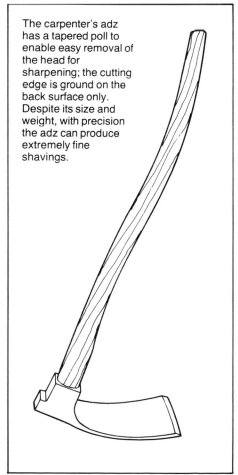

The carpenter's adz has a tapered poll to enable easy removal of the head for sharpening; the cutting edge is ground on the back surface only. Despite its size and weight, with precision the adz can produce extremely fine shavings.

CHAPTER SIX

Boring Tools

*Holes are usually bored
either to house screws, bolts, nails or
dowels, or to serve as a
decoration feature in a piece of
furniture.*

OLES ARE USUALLY BORED either to house screws, bolts, nails or dowels, or to serve as a decorative feature in a piece of furniture. Boring tools can also be used to remove waste-wood in joints, low-relief carvings and sculptures. The type of boring tool to be used for these jobs is determined by the size and purpose of the hole which is being bored.

Boring tools are held in a hand brace, a hand drill or an electric hand drill, the shank end being styled to fit the chucks of these tools.

Brace tang bits are manufactured in several styles, all having a particular purpose. Each bit has to have a screw nose so that it can be drawn into the wood, a spur to cut the periphery of the hole (at the same time cutting the top fibers cleanly) and a cutter, with a twist or spiral, through which the chips can be passed out of the hole while it is being cut. The most popular of this type is the solid center bit.

This has a single twist formed around a solid center, with a stout screw nose and spurs. It is available in many sizes and cuts accurately, leaving a smooth finish, and is excellent for general-purpose boring.

The Jennings pattern bit is the best designed bit, giving a very good hole. It has a double twist to provide greater support in the hole and to help keep the bit square. This is the craftsman's bit, being fast and clean-cutting in the majority of timbers.

The Scotch nose bit was designed specially to cut holes in hardwoods. Like the Jennings bit, it has a double twist, but has a coarser-pitched screw

Producing the machine auger twist.

Hand-Boring Bits

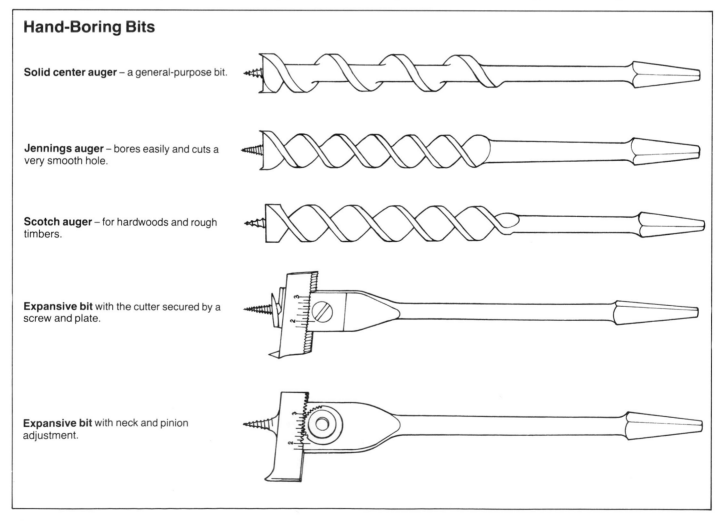

Solid center auger – a general-purpose bit.

Jennings auger – bores easily and cuts a very smooth hole.

Scotch auger – for hardwoods and rough timbers.

Expansive bit with the cutter secured by a screw and plate.

Expansive bit with neck and pinion adjustment.

nose and, instead of spurs which might break off, it has side wings which combine with the cutter to scribe the hole and cut the waste.

To go to the expense of buying a set of bits might seem an extravagance to the woodworker who will use them only very occasionally. The problem can be solved by buying an expansive bit, which comes in a number of styles.

The English pattern is available in two sizes, each having two separate cutters, with additional optional sizes up to 6 inches/150mm in capacity. It comprises a solid screw nose with an adjustable cutter, at the extremity of which is a spur. The cutter can be set at 1/32-inch/0.8-mm divisions against a datum mark on the body, and is held in position by a screw and plate. The expansive bit should be used only to bore shallow holes in softwoods or soft hardwoods.

One American design has the size set with a small pinion which fits in a rack on the underside of the cutter. A datum mark on the body lines up with one on the cutter.

All these bits have a brace tang which fits exactly into the veed jaws of the brace chuck. The brace comes in a variety of styles and sizes, and is plain or ratcheted to allow the brace to be used in confined spaces. The corner brace has both a swinging and a fixed frame to allow it to be used in restricted areas such as near skirting boards. Normally of 10-inch/250-mm sweep, the steel frame is fitted with a hand and chest handle made either of wood or plastic. Where working space is limited or only small holes are being bored, some craftsmen prefer a brace with as small a sweep as 6 inches/150mm.

A recently developed tool has a four-jaw chuck which will take not only the square brace tang but also round shanked bits.

The use of hand electric drills and vertical drill stands is becoming increasingly popular. Most makes have a three-jaw chuck with a capacity up to 1/2 inch/12mm.

This means that many boring tools, previously regarded as exclusive to the larger pedestal- and bench-mounted machine drills, can now be used.

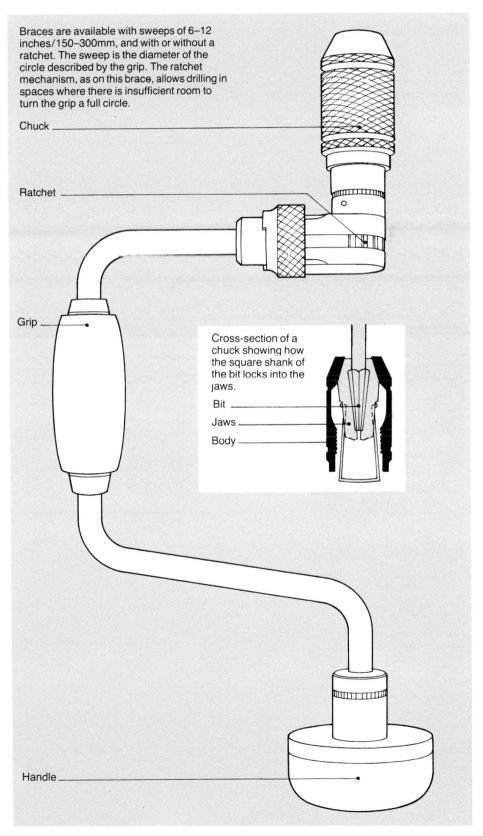

Braces are available with sweeps of 6–12 inches/150–300mm, and with or without a ratchet. The sweep is the diameter of the circle described by the grip. The ratchet mechanism, as on this brace, allows drilling in spaces where there is insufficient room to turn the grip a full circle.

Chuck

Ratchet

Grip

Cross-section of a chuck showing how the square shank of the bit locks into the jaws.

Bit
Jaws
Body

Handle

Machine Bits

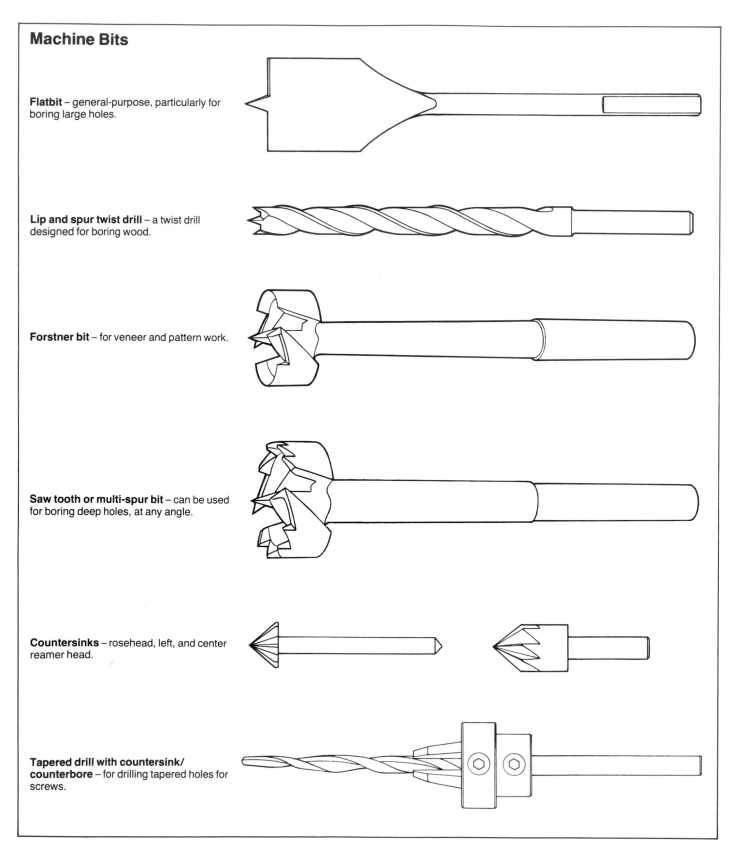

Flatbit – general-purpose, particularly for boring large holes.

Lip and spur twist drill – a twist drill designed for boring wood.

Forstner bit – for veneer and pattern work.

Saw tooth or multi-spur bit – can be used for boring deep holes, at any angle.

Countersinks – rosehead, left, and center reamer head.

Tapered drill with countersink/ counterbore – for drilling tapered holes for screws.

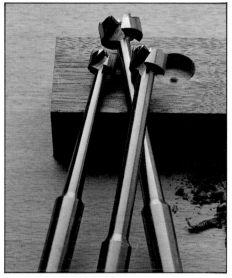

Multi-spur bits with two clearing slots.

The most versatile bit designed for the hand electric drill is the spade, or flat, bit. It has flats on the shank for positive grip in the drill chuck. The long brad point allows holes to be bored in soft- or hardwood, end grain or knotted timbers and wet or stringy timbers. The bit can also cut holes in perspex and other plastics.

Always allow the bit to stop before withdrawing it from the hole, and take care when boring through holes, again allowing the bit to stop before withdrawing. An extension shank can be fitted to increase the reach, which is useful particularly when boring holes in almost inaccessible places. The bits range in size from ¼ inch/6mm to 1½ inches/37mm, and are invaluable not only to the wood craftsman and turner,

but also the electrician and plumber.

When he wants to bore small holes, the wood craftsman often uses jobber's drills designed for use on metal. These bits do not have a point for accurate positioning of the hole and have inadequate chip clearance. As a result, they become burnt and, indeed, rarely bore a round hole in the right place. A bit designed to cut wood smoothly, accurately and without burning is the brad point, or lip and spur drill. It has a round shank and is made of high-speed steel, in sizes ranging from ⅛–½ inch × 1/16 inch/3–12mm × 1.5mm. Positive centering is assured, the twist provides adequate chip clearance and the breakthrough is quite clean. Treated properly, it will give a lifetime of use.

A Forstner bit is a very useful tool,

To hide screw heads in a construction, plugs are drilled in a piece of matching wood and then dropped into the pre-bored hole.

Doweling Jigs

The simplest method of aligning a doweling joint is to use a dowel locator, **1**. Having first positioned the locator on one of the halves of the joint, the other half is aligned over it. By pressing down, a clear mark will be made on the timber; a collar prevents the locator being pushed all the way into one of the halves. Having marked the doweling joint, a doweling jig will be required to drill the dowel holes square and true in the two halves of the joint. Both the doweling jigs, **below**, have brushes which keep the drill bit vertical during drilling. The jig, **2**, has five brushes from ¼–½ inch/6–12mm and the fences open and close simultaneously, so the bit is automatically centered. With the jig **3**, the bush is held by a clamp. After the jig is attached to the timber to be drilled, the position of the bush is then adjusted separately. Bushes range in size from ³/₁₆–½ inch/4.5–12mm.

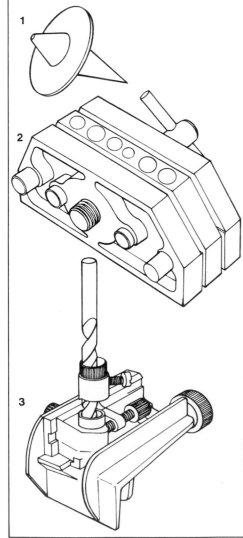

Left and above, a more versatile doweling jig will both guide the drill bit and align the holes in both halves of the joint. There are two heads: one is a reference or fixed head, while the position of the other is adjustable. Two bush carriers hold the drill bushes and adjustable side fences. When drilling the end grain of one half of a doweling joint, left, the adjustable head is positioned just clear of the work and the jig is then held in place by one of the clamping screws. The side fences must be touching a side face of the work. To drill the side edge of the other half of the joint, the adjustable head is removed and the jig is inverted; it will then be perfectly aligned for drilling the other half of the joint. When drilling boards longer than the slide rods, above, remove both the heads and press the side fences hard against the board. A dowel placed in the previously-drilled hole will ensure that the holes are evenly-spaced along the edge. Invert the jig to drill the holes in the other board.

designed for boring flat-bottomed, shallow holes in any timber. Although it has a tiny brad point, which is needed for accurate centering, the bit runs on its periphery. It has two cutters and a ½-inch/12-mm shank, and bores full, overlapping or off-the-edge holes in either face or end grain. Its complete accuracy makes it a good tool in veneer and pattern work. Use it at fairly slow speeds and a fast rate of feed to avoid any possible burning of the rim.

A bit very similar in basic construction to the Forstner bit is the multi-spur, or saw tooth, cutter. Instead of the rim, a series of rip saw-like teeth cut the periphery of the hole before the cutters come into contact with the stock. With a ½-inch/12-mm shank, the sizes range from ⅜–3½ inches/10–87mm wide. The bit bores cleanly in timber hard or soft, wet or dry, stringy or knotted.

The base on a drill guide slides on rods so that holes can be drilled at any angle; an adjustable stop on one rod sets the depth of the hole.

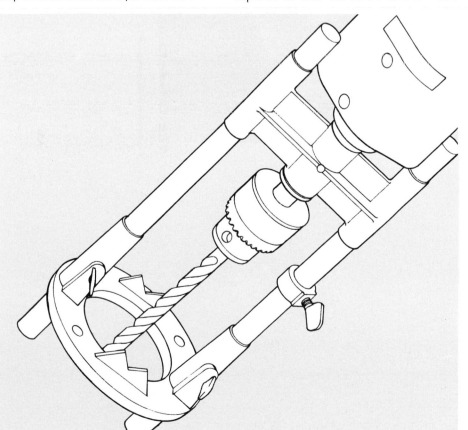

Unlike the Forstner bit, it can be used to bore deep holes even in end grain, and is, as a result, often used by the woodturner for boring boxes, vases and other woodturnings. The American pattern has only one cutter instead of the two in the English version.

No kit would be complete without a countersink bit, which cuts recesses in drilled holes to receive the head of a countersunk screw. There are two types of head – the rosehead, which has a conical cutting end, and is the most common, and the center reamer, which has a flattened V-shaped head and is used to countersink hinges. They are made of carbon and high-speed steel and are available in several sizes. The bit should never become clogged, or it will burn.

Another special boring tool has been developed from the countersink and counterboring machine bit, for use particularly in the hand electric drill. This is tapered to suit the screw and is fitted with a counterbore and countersunk collar which acts as a depth stop. This can be fixed also in any position along the drill by screwing down two set screws. This tool will save time if a great deal of this work is to be done.

Inserting screws is often a difficult process, and is especially so when the screws are excessively long. Usually, this means the timber has to be pre-bored, using two sizes of bit, which is a lengthy and laborious business. To avoid this, extra long taper point drills can be used. These are extremely efficient, are made in sizes related to screw gauges and have a reach of approximately 6 inches/150mm.

Many craftsmen, in using screws or even bolts in a particular construction, object to the sight of the steel heads in the wood. To improve the finish, the holes can be pre-bored so that the screw heads drop below the surface, and then a plug of the same wood is glued in place to hide the screw completely. There are several types of plug cutter, but the best is one which cuts a plug that is perfectly smooth and of the correct size, and which is ejected through an escapement at the side of the bit. Plug cutters are available from ⅜–1½ inches/10–37mm wide. They are beautifully-made tools which are capable of cutting parallel plugs from end or face grain in hard or soft wood.

A simpler kind of plug bit will cut thin tapered dowels 2¼ inches/56mm long, and in a variety of widths from ⅜–1½ inches/10–37mm. It has a ½-inch/12-mm shank.

The smaller sizes of all these machine bits can be used in a hand drill.

The drill guide is a useful tool when doing straight or angled boring to a pre-set depth using brad point bits, spade bits or an engineer's drill.

This guide can be used with electric drills of ¼-inch/6-mm or ⅜-inch/10mm capacity and a ⅜-inch threaded shaft, with a motor housing not wider than 3 inches/75mm. The base of the unit slides on two rods and can be accurately angled using a set square. This jig ensures accurate on-the-spot boring as well as the exact centering of round stock.

CHAPTER SEVEN

Chisels

*The chisel
is one of the most widely
used and generally
abused tools currently
made.*

THE CHISEL IS ONE of the most widely-used and generally abused tools currently made. Although available in a wide variety of shapes and sizes, each designed for a different purpose, many woodworkers use only one type of chisel for every job. However, many modern manufacturers have rationalized their ranges to the extent that many special types of chisels are hard to find, so often the woodworker has to make do with what is available.

The chisel performs many functions and, depending on the job, is used in either the left or right hand. Often, the chisel is in use for long periods at a time, and may even be used with a mallet.

Bearing these facts in mind, the chisel must be selected not only for the quality of its steel and the manufacturer's assurance that it has been properly heat-treated, but also for the excellence of its handle design. Too often in the past, chisel handles were made to suit the blade and shape of the woodturning machine, and not of the human hand. Many of these chisels were not only difficult to hold but, with prolonged use, made the hand sore and the arm ache. Fortunately, the present-day manufacturer has paid greater attention to ergonomics, and chisels are designed with not only the hand in mind but also the material used, so that it is comfortable to grip. Many of these handles are practically indestructible, and certainly, most are splitproof. The overall design and color has also been considered.

When looking for a chisel, the woodworker would be well advised to seek out those from a reputable manufacturer. Hold the chisel in both hands, check the handle for firmness and the blade for flatness. At the same time, notice the quality of the grinding, as a badly-ground chisel will require a great deal of work on the oilstone before the fine hairline cutting edge is produced. If a bevel edge chisel is being selected, make sure that the bevels are finely ground. Whilst extreme accuracy of size is not vital, usually if a mortice chisel is bought, it is best to check its accuracy.

Examine the chisel closely to check the alignment of the blade and the handle. Balance should not be ignored,

Chisel and Gouge Types

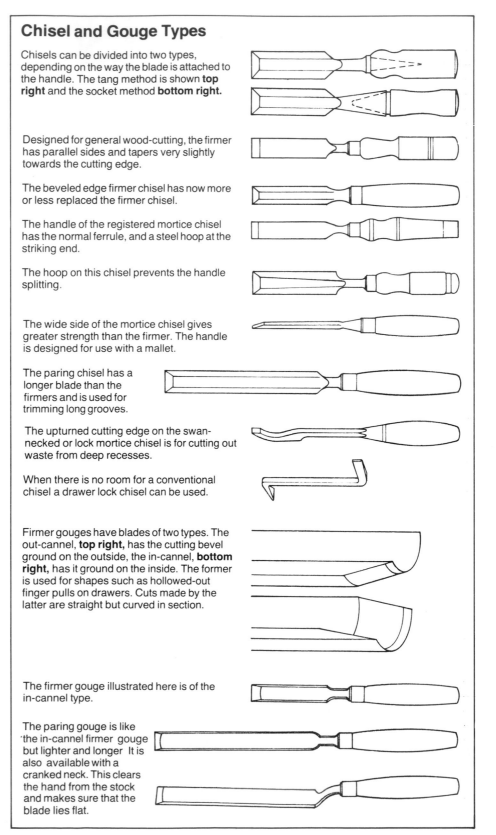

Chisels can be divided into two types, depending on the way the blade is attached to the handle. The tang method is shown **top right** and the socket method **bottom right.**

Designed for general wood-cutting, the firmer has parallel sides and tapers very slightly towards the cutting edge.

The beveled edge firmer chisel has now more or less replaced the firmer chisel.

The handle of the registered mortice chisel has the normal ferrule, and a steel hoop at the striking end.

The hoop on this chisel prevents the handle splitting.

The wide side of the mortice chisel gives greater strength than the firmer. The handle is designed for use with a mallet.

The paring chisel has a longer blade than the firmers and is used for trimming long grooves.

The upturned cutting edge on the swan-necked or lock mortice chisel is for cutting out waste from deep recesses.

When there is no room for a conventional chisel a drawer lock chisel can be used.

Firmer gouges have blades of two types. The out-cannel, **top right,** has the cutting bevel ground on the outside, the in-cannel, **bottom right,** has it ground on the inside. The former is used for shapes such as hollowed-out finger pulls on drawers. Cuts made by the latter are straight but curved in section.

The firmer gouge illustrated here is of the in-cannel type.

The paring gouge is like the in-cannel firmer gouge but lighter and longer It is also available with a cranked neck. This clears the hand from the stock and makes sure that the blade lies flat.

as a thick, heavy-bladed chisel will be out of balance and make working difficult. If the handle is wooden, check it for splits and other flaws. The quality of steel or the correctness of its heat treatment cannot be checked, so once again, emphasis must be laid on the selection of brand name tools which invariably carry a guarantee. A good chisel, correctly sharpened, will keep its edge for a long time, unless it is abused, so that valuable time is not wasted in resharpening fairly new chisels.

Chisels can be classified under three groups – general-purpose, mortice and special. The handles of all these chisels are secured either by a tang or socket, and are made from either wood or plastic. The most popular wood amongst British craftsmen is European box, but ash and beech are also widely used, although one British manufacturer uses jarrah and another, rosewood.

The handles must be checked in the knowledge that they may be struck with a mallet and that often the chisel may be wrenched from side to side. Therefore, the wood must be close-grained and free from knots, and must not be prone to surface cracking and splitting. A hard but pliable wood, which is shock-resistant, is essential. Few of the available timbers have all these features – ash, for example, is fairly hard, pliant and knot-free, but its grain is quite open. Beech is closer-grained but not quite as pliant.

The early English edge-toolmakers chose boxwood, which is probably unequalled in its suitability. Boxwood (*Buxus sempervirens*) is indigenous to Europe and is regarded largely as a decorative shrub or tree for hedges and gardens. It has a narrow trunk and grows wildly in misshapen branches, reaching about 8 feet/2.4m high. The handles are turned from the branches.

Box was used by the printer and engraver in the past, and it is likely that these blocks came from trees of some size. The blocks were cut with the grain running vertically, and sold by the cubic inch. Various substitutes for box have been tried, such as San Domingo boxwood and zapatero, but neither of them

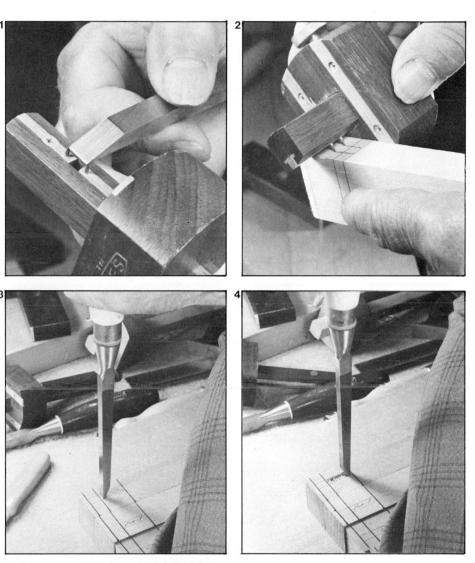

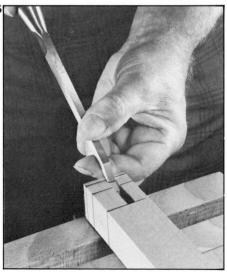

The stages of cutting a mortice for a tenon with secret haunch are illustrated **above and left: 1** The mortice gauge is set using the chisel itself rather than a ruler, as this is more accurate. **2** The wood is marked. **3–4** The chisel works along the projected mortice area. **5** The haunch is cut.

are equal to the European box. The timber is butter-yellow in color, although the smaller timbers have a grey streak running through them.

The splitproof plastics are cellulose acetate butyrate, cellulose acetate and polypropylene.

Half a century ago, the most widely-used chisel was the firmer. This has a thick steel blade with handles of box, beech, ash and plastic. It is a general-purpose tool, suitable for most jobs and, as it has a stout blade, can be driven with a mallet if the handle is wooden, and with a hammer if it is plastic. However, the popularity of the firmer chisel has been superseded by that of the bevel edge, as it is virtually identical, and woodworkers are reluctant to buy two chisels which perform the same tasks.

The bevel chisel was designed to cut dovetails – the long edges are beveled to cut the corners of the dovetails. When buying this chisel, the bevel must be checked for a fine edge, as some manufacturers produce chisels with a bevel edge that is far too thick.

All fine cutting can be carried out with the bevel chisel but it should not be used for deep cutting. Horizontal and vertical paring, chamfering both with and across the grain, dovetailing, housing, stopped housing and dovetailed housings across narrow boards are all well within the capacity of this chisel.

The cutting of mortices is a tough task for a chisel, but luckily a chisel has been produced to combat this. The London pattern sash mortice chisel has a boxwood handle and a thickened blade which, together with a solid bolster, gives great strength. A leather washer fitted between the brass ferrule and the bolster shoulder absorbs shock. The blades have a reach of 6½ inches/162mm and are accurate to width, although it is advisable to set the mortice gauge spurs to the chisel and ignore the indicated size.

Illustrated **top left** on the **facing page** are heavy-duty double-hooped chisels. The handles are ash, with steel hoops at both ends to prevent splitting. **Bottom right** are bevel edge chisels with boxwood handles. **Inset:** double-hooped mortice chisels with ash handles and leather shock washers.

A variation on this chisel is one which has a purpose-designed plastic handle, and a nylon washer. This handle is ideal for the mortice chisel. Its size suits the majority of hands, the four-square cross-sectional shape gives positive grip and helps to counter any slipperiness experienced with the material. Also, it is curved to avoid blistering the hands.

Another chisel which can be used in morticing is the steel hooped chisel, known as the registered mortice chisel in Britain. This chisel is of normal length, with a blade thicker than the firmer, but its special feature is a steel ferrule and steel hoop at the end of an ash handle.

A similar chisel has an oval handle, but the timber extends beyond the steel hoop, so that it can be driven with a mallet only.

A long, thin paring chisel is needed to cut housings (slots across the grain). The blade is thin and beveled on its long edges and has the required reach, as well as the clearance afforded by the bevel. This chisel is the only one which can be used to cut a dovetailed housing.

It is available with a cranked, or trowel, neck, which is an excellent feature as it clears the hand from the stock, and ensures the blade lies flat, particularly on very wide work. When selecting this type of chisel, it is vital to check that the chisel is flat across both the width and length. Although slight up-turning of the blade is an advantage, the opposite would make flat cutting impossible.

When paring vertically with a chisel or paring gouge, it is best to work on a cutting board to protect the bench top. One can be made easily from a piece of beech or other even-grained hardwood. If a strip of wood is screwed onto the underside of the board, it can be held in a vice when in use.

A popular chisel amongst some workers is the butt, or pocket, chisel. With the exception of a shorter blade, this chisel is identical to the bevel edge chisel. It is very useful when cutting the housings for butt door hinges.

The swan-neck mortice chisel is used to hollow out the mortices for door locks and other deep recesses. It has a beech handle and a long blade with an up-turned cutting edge, made of steel.

Another useful tool is the drawer-lock chisel, which cuts lock recesses in drawers and other confined spaces. It has a square-sectioned steel bar with both ends cranked at right-angles, and ground to a fine edge. Once in position, the chisel is struck with a hammer.

Gouges

Gouges which are classified as in-cannel and out-cannel are included in the special chisel group. Cannel is a word which does not appear in the English dictionary but, nevertheless, is a term well-known in both the Sheffield chisel industry and in North America.

In-cannel means the cutting bevel has been ground on the inside, and out-cannel means it has been ground on the outside. In- and out-cannel gouges are available as firmer and paring gouges, in a variety of blade sizes and curves. In-cannel gouges with longer blades and cranked necks are also on the market. High-quality gouges have boxwood handles.

The striking of any chisel or gouge, other than the registered pattern or drawer-lock chisel, should be done with the joiner's mallet. This has a handle of pliant ash, the head is best made from Danish beech, and the handle socket is tapered to ensure a tight fit.

Knives

Another edge tool which is vital in the workshop is the knife, because it is needed to mark out lines which have to be cut either with the saw or chisel. The woodworker can choose between knives with blades ground on one side only, or on both, and with wooden or plastic handles. Knives needed for other work are available in various shapes. The best-known craft knives were developed in Sweden many years ago when woodcraft was first taught in schools. They are known as Sloyd knives, come in a variety of blade lengths and can be used for many workshop tasks as well as for whittling. Knives designed for whittling and chip carving will be discussed in detail in Chapter Ten on carving tools.

CHAPTER EIGHT

Abrasives

*Abrasive tools are
cutting tools with either sharp edges of
steel or various grades of natural
or man-made grit*

Most of the tools looked at in the previous chapters have been primarily traditional in their design and use. However, this does not apply as much to abrasive tools, as a number of new and exciting developments have occurred during the past two decades.

Abrasive tools are generally regarded as finishing tools, although some are used for basic work. All are cutting tools with either sharp edges of steel or various grades of natural or man-made grit to fashion the surface of the wood.

The earliest abrading tool was the rasp, still in use today, made of flat, half-round or round steel with individual teeth cut with a punch. Rasps are not supplied with handles, but it is dangerous to use them without one, and the tang of the rasp can be fitted into a ferruled handle which is sold separately.

Rasps were originally made by hand, each tooth being punched out separately, but now they are cut with machines.

The rasp most popular today is the cabinet rasp, usually half-round, and from 10–12 inches/250–300mm long. This rasp is very good for the fast removal of wood, although the resultant surface has to be smoothed off.

There are four recognized degrees of coarseness in rasps and files – rasps are graded coarse, bastard cut, second cut and smooth, and files are graded wood rasp, cabinet rasp bastard, cabinet rasp second cut and cabinet rasp smooth.

A file cuts more smoothly, although it becomes clogged with waste quickly. Its cutting edges are formed by machining at an angle across the face of the file.

The single cut file for smoothing has teeth lying in one direction, but the double cut, which is rougher, has teeth lying across the file in two directions.

The cabinet file used by the woodworker is flat or half-round, and is tapered. It is available in smooth, second and bastard cut.

A riffler file must be used when filing curved surfaces, or areas which are difficult to reach. Although this is used mainly by the carver and sculptor, the woodworker will find it useful too. Rifflers are discussed in more detail in Chapter Ten, on carving tools.

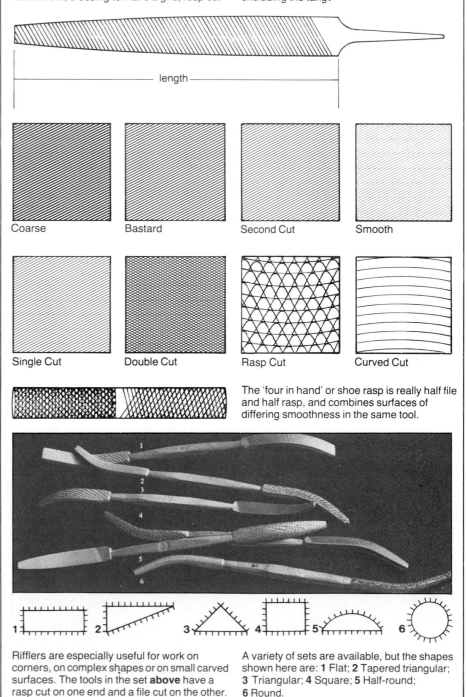

Files, Rasps and Rifflers

Rasp and wood files are essential for smoothing edges and shaping small areas. They are classified according to the arrangement of their teeth. These can be single-cut, running in diagonal parallel lines, double-cut, crossing to make a grid, rasp cut and curved cut. There are four grades of smoothness: coarse, bastard, second cut and smooth. However the number of teeth per inch will vary in proportion with the length of the file. The length of these tools is measured excluding the tang.

length

Coarse Bastard Second Cut Smooth

Single Cut Double Cut Rasp Cut Curved Cut

The 'four in hand' or shoe rasp is really half file and half rasp, and combines surfaces of differing smoothness in the same tool.

Rifflers are especially useful for work on corners, on complex shapes or on small carved surfaces. The tools in the set **above** have a rasp cut on one end and a file cut on the other.

A variety of sets are available, but the shapes shown here are: **1** Flat; **2** Tapered triangular; **3** Triangular; **4** Square; **5** Half-round; **6** Round.

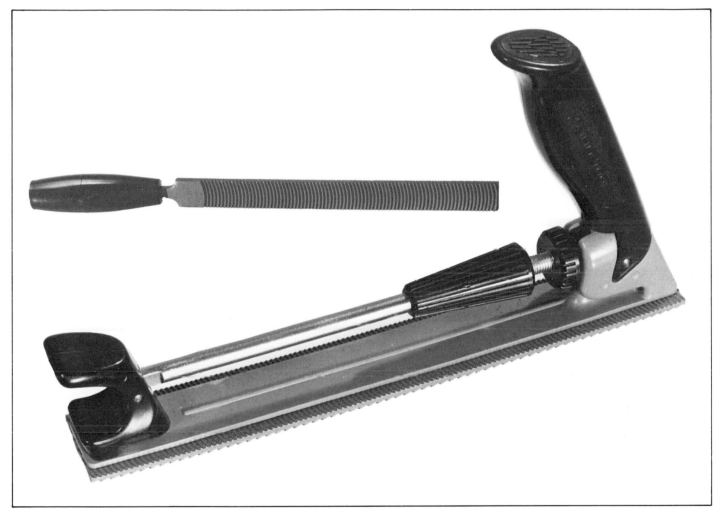

A very useful tool, similar to the metalworker's dreadnought file, is the Aven Filemaster. It serves two purposes, as the curved teeth on one side give an almost plane-like shaving, leaving a superb finish, and the cross-filed teeth on the opposite side can be used to rough down. It is 10 inches/250mm long, and is available in both flat and half-round styles.

The Stanley Surform has become increasingly popular in recent years. This unique tool was patented by an English firm, Firth Cleveland, almost twenty years ago and, although the tool forms have changed, the original blade design remains.

The blade is pierced to form rows of cutting edges, which leave a surface similar to that of the rasp. The volume of wood removed is considerable, and

The Aven Filemaster, **top,** has one surface of cross-filed teeth for rough work and one of curved teeth, which gives an extra-smooth finish. The Trimmatool, **bottom,** offers the special facility of an adjustable curved blade for concave cuts. The curved teeth of these tools, **left,** are virtually self-clearing.

Surform tools are used for forming and trimming wood, plastics and soft metals. They can enlarge round holes or finish curved work. Basically, a Surform is a hollow rasp. Sharp-edged holes perforate its steel blades so that wood is removed quickly and efficiently. The major types of Surform are illustrated on these pages. The plane type is shown in use, **above. Right,** a half-round blade with attached easygrips. On the **facing page,** from the top, are a Surform plane, a ripping plane and a round file.

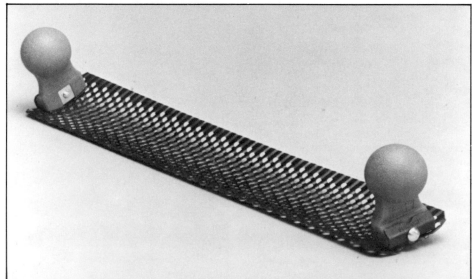

the blades can be flat, half-round or round, although the flat one is flexible enough to follow a curved surface. A large variety of blades is available, all of which can be fitted with handles to convert the tool into planes and files. The pierced blade allows the shavings to pass through into the body channel. It is a first-class roughing tool.

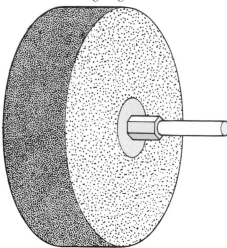

An abrasive drum.

A newer set of tools are those that use abrasive papers or cloth. One of the least expensive is the abrasive drum. This consists of a cylinder of foam plastic or rubber with a central arbor, or shaft, usually ¼ inch/6mm in diameter, which is mounted in an electric drill or flexible shaft. An abrasive band is placed over the drum to make an ideal tool for cleaning up both flat and curved work.

A better-lasting tool is the abrasive flap wheel. This consists of a central plastic core to which is attached a number of flaps of abrasive paper or cloth. The wheel has either a fixed or removable shaft, which is screwed in, and both types fit the chuck of the electric drill. Larger flap wheels can be substituted for the wheels on the bench grinder.

The flaps are coated in man-made grit in the usual abrasive grit sizes between 40 and 320. The wheel used in the workshop is from 1½–3 inches/37–75mm in diameter. They are so designed that, as the flaps wear down, new grit is exposed. Never push the wheel, but let it rotate at its own speed.

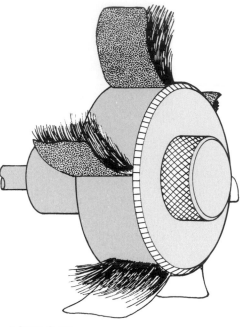

A Sand-O-Flex.

A wheel with replaceable abrasive is marketed under the name Sand-O-Flex. This tool consists of a metal body, with a number of slots around the periphery. Abrasive cloth strips are fed through the slots from a center boss and are cushioned by backing brushes. This cushioning allows the abrasive cloth to work into, around, and over any shape. As the cloth wears, more can be fed out by turning the outer cover. Several sizes are available, with a reserve coil of a 60–150 inch/1500–3750mm capacity.

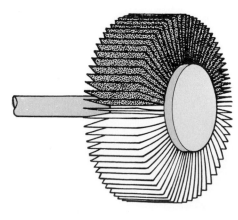

An abrasive flap wheel.

The Sand-O-Flex can be used with an electric drill or fitted to a stationary power source. The full range of abrasive grit cloths is available for use with this wheel. This tool is essential for the wood sculptor, because it will not groove or burn.

Always wear safety goggles to protect the eyes from tiny pieces of flying wood when using tools of this type.

The abrasive disk is neither as useful nor as safe as the flap wheel. There are a number of flexible backing disks available, against which the abrasive disk must be placed and which are suitable for mounting in an electric drill. Great care must be exercised when using this tool on wood, because the edges of the disk can badly score the surface.

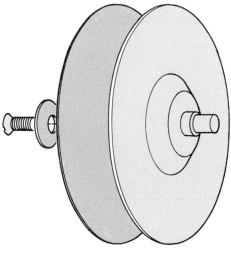

An abrasive disk with rubber backing.

Abrasive papers come in many forms and grits and in a fairly wide range of sizes. The most commonly-used abrasive paper is flint paper, obtained in standard-size sheets of 11⅛ × 9⅛ inches/280×230mm. The flint is glued to strong paper and will produce a nicely finished surface quickly, if several grades from coarse to smooth are used. Unfortunately, the flint wears quite quickly and the paper rapidly clogs.

A better paper with a slightly longer life is garnet paper. This is natural garnet which has been crushed to provide a cutting grit.

The best paper is surfaced with aluminum oxide, which is a tough, hardwearing artificial grit generally used in industry for working hardwoods on large machines. Used in the home workshop, aluminum oxide papers will give prolonged use and can be cleaned.

Wet-and-dry paper is growing in popularity. The grit used on these papers is silicon carbide, produced by heating quartz sand. The paper and glue are waterproof, so the paper can be used with a lubricant. When used wet, the paper will not clog as easily and consequently dust is reduced.

The most effective of grits is tungsten carbide but, unfortunately, this is not yet widely available. The grit is fixed to very fine wire mesh backing, and can be cleaned. As a cutting material, it is about one hundred times more efficient than flint paper.

Papers used in the hand should always be backed with cork. Cork blocks are available for this purpose and are superior to wooden blocks, as they are soft and slightly flexible. If metal sheets are used, they should be attached to a strip of wood.

Steel wool is a mild abrasive, and is available in several sizes. It is useful particularly when cutting back polishes on the lathe and can also be used on other curved work.

The lives of belts, disks and flap wheels can be extended by using an abrasive restorer. This is sold in stick form and is held against the moving belt to remove the dust.

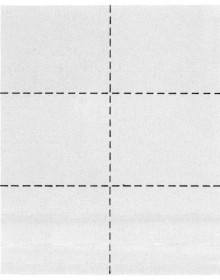

Abrasive paper should be divided as shown.

Abrasive Papers

The table below shows the different sorts of abrasive paper available, in terms of fineness and type of material.
For approximate equivalents in the different numbering systems read vertically down the columns.

	Very fine Polish and smooth between finishing coats. Smooth after final coat.	**Fine** Final surface work before application of sealer.	**Medium** Intermediate removal of roughness prior to fine work.	**Coarse** Initial sanding on most woods.	**Very coarse** Extremely rough work, often in place of rasp or file.
Aluminum oxide *Use* Hardwood and soft metals, ivory, plastic.	600–220	180–120	100–80	60–36	30–12
Silicon carbide *Use* All materials including glass and cast iron. Wet sanding. Final finish.	600–220	180–120	100–80	60–36	30–12
Garnet *Use* Sanding hard or soft woods, plastic, soft materials like horn.	10/0–6/0	5/0–3/0	2/0–1/0	½–1½	2–4½
Flint *Use* Rough sanding, removing old paint and finishes.		00–1	1½–M2	S2–3	
Tungsten carbide *Use* A very efficient cutting material.		Fine	Medium	Coarse	

CHAPTER NINE

Turning Tools

*Woodturning
is one of the most enjoyable of
the woodcrafts; it
allows the woodworker to be
completely creative*

Woodturning is one of the most enjoyable of the woodcrafts. Every woodworker will at some time feel the fascination of woodturning. Certainly, it is one craft which allows the woodworker to be completely creative, using both hard and soft timber. The versatility of a lathe means small and large timber can be used, giving the woodworker the chance to experiment with rare and exotic timbers, and also local timbers from the garden and hedgerow, which are rarely of any size but are often most beautiful. You will begin to appreciate wood as you work with it, and when you see the ribboned shavings flying, you will experience the thrill and pleasure of creative woodturning, and be well on the way to producing an infinite variety of beautiful objects.

When considering embarking upon this new craft, the two factors most likely to deter the would-be turner are his lack of experience in woodturning, and the need to buy a lathe.

Choosing a Lathe

Many excellent lathes are manufactured in both Europe and North America, and one can be found to satisfy both need and pocket. Woodturning lathes fall into three groups – heavy lathes designed for production work, lighter but sturdily-built lathes designed for schools and colleges, and lathes designed for the home craftsman – some with attachments which convert them into multi-purpose machines, combining circular sawing, band sawing, planing and sanding. When choosing a lathe, the craftsman must consider for what the lathe will be used, for how long, and the amount of space available. If the craftsman does not anticipate turning much long work, then a short bed lathe would be suitable. However, the larger lathe does have advantages over the light amateur type, because it is sturdier, and more able to bear the heavy vibrations sometimes produced.

Apart from the small lathe attachments supplied with power drills, all woodturning lathes have belt drives, using either vee or flat belts. These are designed to provide up to four different

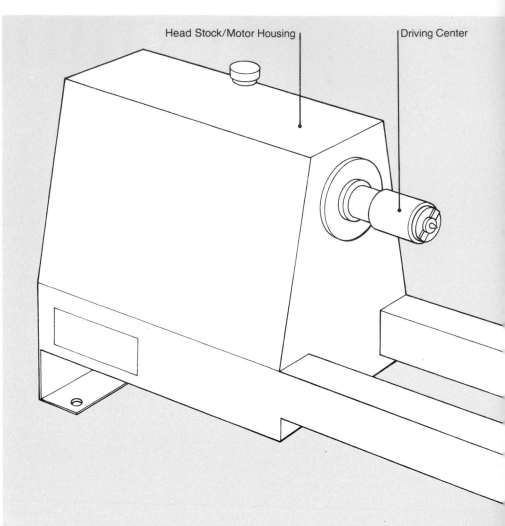

Head Stock/Motor Housing   Driving Center

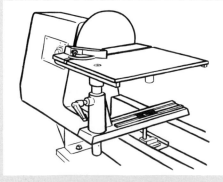

The disk sander is an extremely useful accessory. It is shown here assembled for use in conjunction with a table and mitre guide.

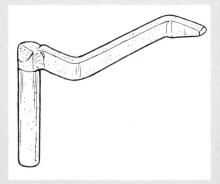

This tool rest is especially shaped for use in the making of bowls.

The lathe provides for two types of turning. In between centers turning, the wood is held between a moveable tail stock which has an adjustable spindle and a fixed head stock which houses the motor which powers the driven spindle. A tool rest is located between these centers. In face plate turning the wood is attached to the face plate which is then attached to the driven spindle.

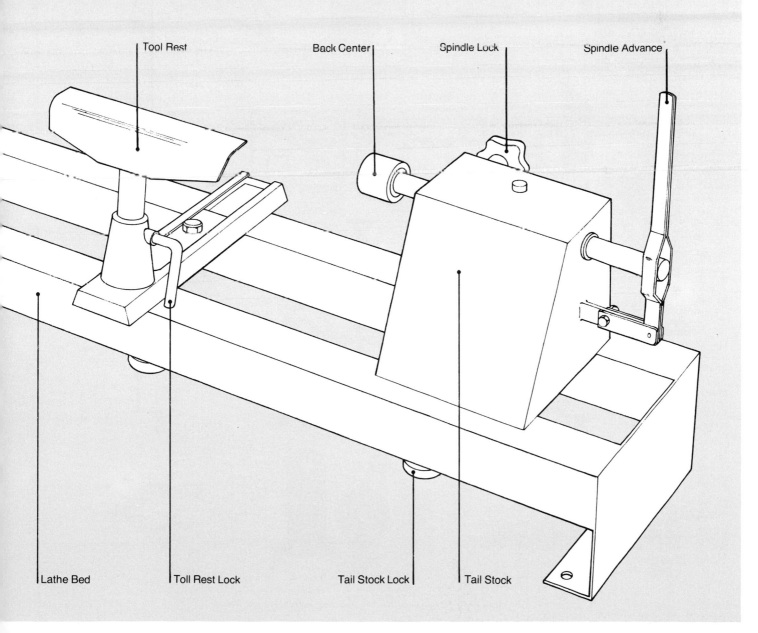

Tool Rest   Back Center   Spindle Lock   Spindle Advance

Lathe Bed   Toll Rest Lock   Tail Stock Lock   Tail Stock

LATHE.

by Mr. H. Maudslay.

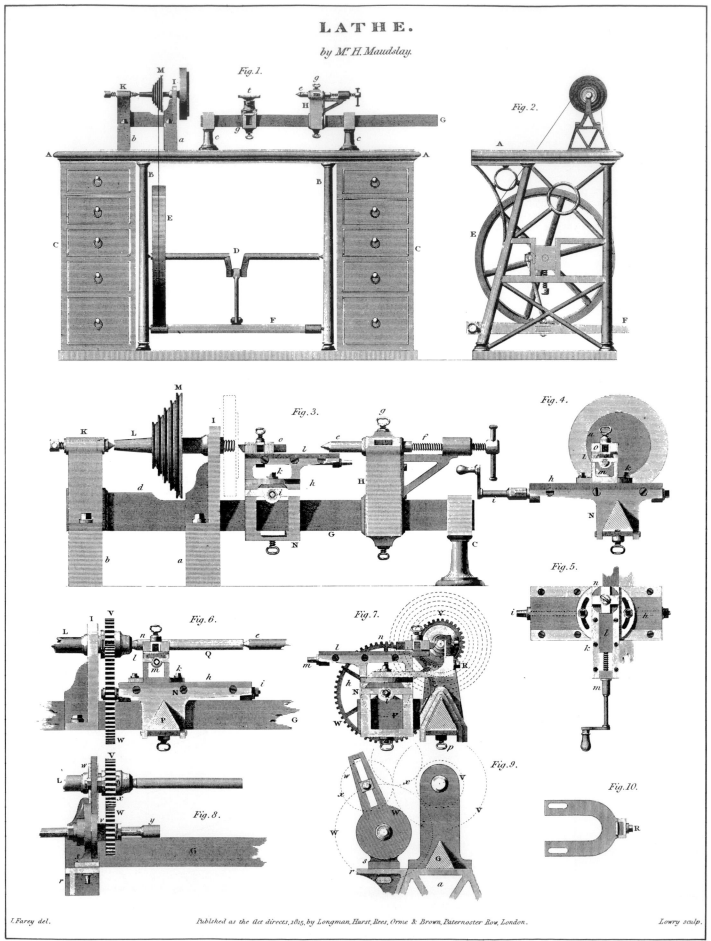

I.Farey del.   Publshed as the Act directs, 1815, by Longman, Hurst, Rees, Orme & Brown, Paternoster Row, London.   Lowry sculp.

It would be idle to suggest that modern lathes are no more sophisticated than their ancestors, but the lathe from the early nineteenth century, **facing page,** shows that basic principles have remained the same. The multi-purpose machine, **above,** combines the functions of a lathe with those of a circular saw, a morticer, a planer and a sander.

speeds, ranging from roughly 500 rpm to 3000 rpm. Belt-changing devices vary and some are most difficult to operate. If money is no object, a lathe having a variable-speed drive, and operated by a lever or wheel, should be considered. The final selection must be made in the knowledge that the lathe is supplied with a number of useful work-holding attachments, and that it can be positioned in the workshop with sufficient room for the operator to work in comfort and safety.

Woodturning can be divided into two distinct work areas. When work is turned between centers, it is mounted between the head stock drive and the tail stock, and this is known as spindle turning. In face plate work, blocks are held either by the face plate or by various chucks, and are mounted on the head stock mandrel. Many lathes have a head stock spindle screwed at both ends, enabling face plate work to be carried out on the outboard end, so that work of a larger diameter can be executed. One particular lathe has a rotating head which makes the assembly of larger work possible, and eliminates left- and right-hand face plates.

The lathe selected must have an electric motor of adequate horsepower and be phased to suit the workshop supply. Any advice offered by the lathe supplier should be taken. The lathe must also be fitted with a switch of an approved type, positioned conveniently close to the left hand on the head stock.

Lathe Equipment

This page shows most of the attachments available for your lathe: **1** A face plate with screws for attaching to the work. Screw holes can be avoided by gluing an intervening piece of wood to the work, with a piece of paper sandwiched in between to make later separation possible. **2** A screw chuck. **3** A cup chuck. **4** The Coronet cone driving center. **5** A child coil chuck and timber ready for assembly. **6** A collar chuck, for longer timber. **7** A collet chuck, gripping the timber to be worked on. **8** The recently developed expanding collet chuck combines the function of all these attachments. **9** It is ready for the inside turning of a bowl. **10** It is adapted for use as a collar chuck with a splitring. **11** The faceplate is adapted for use as a spigot chuck. **12** Various sizes of mandrels, used for making wheels.
13 Timber bored and assembled on the mandrel ready for wheel-making.

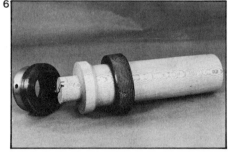

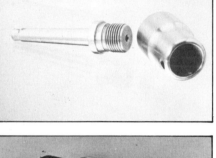
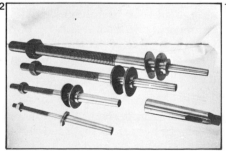
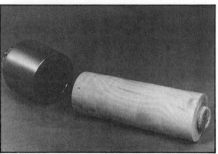

The copying attachment solves the problem of how to reproduce exactly similar patterns on such items as table and chair legs. Each of the adjustable fingers marks a particular part of the work. When the pre-set diameter is reached, the fingers drop out of use.

Lathe Equipment

Most lathes are supplied with a number of standard components. The driving center is either of the two- or four-chisel style, with a morse taper to suit the mandrel of the lathe.

The dead center, either solid or cup, also has a morse taper to suit the tail stock.

The perfect center is one which rotates (often called a running center), eliminating the need for lubricating grease or oil.

One or two face plates are provided, usually with a larger one for the outboard position. There is, however, little need for large face plates, and 3-inch/75-mm and 4-inch/100-mm wide plates will be ample.

Blocks for bowls, boxes and dishes can be held using the screw chuck.

This is also particularly useful for disk work and in the making of picture frames.

The collar chuck, of which there are several variations, is used mainly for holding longer timber, and is ideally suited to the turning of egg cups, napkin rings, slim vases and boxes. There is no possibility of the wood separating from the chuck, and holding holes are eliminated.

One of the most significant advances in lathe design in recent years is the expanding collet chuck, designed and made by an English company – Craft Supplies. This chuck eliminates the need for face plates or other chucks. The expanding collet means the turning is held in a dovetailed recess cut in the base. The chuck can be used as a collar chuck, when a split washer is fitted into a groove cut in the timber which has been previously turned between centers. Long timber of any diameter within the capacity of the lathe can be accommodated. The back plate can serve as a face plate or a spigot chuck. A plate is also supplied to convert the chuck into a screw chuck.

A recent addition to the woodturner's equipment are ground carbon steel mandrels. All have No 1 morse taper arbors and are centered at the other end to receive a revolving center. Each mandrel is part-threaded and fitted with washer and nut. These are ideal for turning batches of toy wheels, rings and pepper mills and have a great number of other uses.

One of the great problems of the woodturner arises when he has to repeat an exact shape. This occurs most frequently in spindle work when making table and chair legs. The latest device created to simplify this is the craft copying attachment, made by Craft Supplies.

It consists of a long tube which is screwed to the lathe base. Adjustable fingers are placed along the tube, each one marking a particular measured part of the work. Each is set up to an exact diameter, the moving part of the finger dropping when the size is reached.

A completely new concept in the holding and driving of work between centers is the cone center, made by Coronet. Here, the driving fork in the headstock has been substituted by a cone which is fitted with a morse taper or, in the case of certain machines, can be screwed on the lathe mandrel. The square section timber is inserted into, and driven by, the cone. Cones with ¼–2½ inch/6–62mm capacities are available.

A similar design which runs on a ball race is available for the tailstock.

The squared timber can be accommodated without end sawing or center popping. Round stock can also be accommodated and if at any time the rounded job is taken out of the lathe, it can be replaced easily without having to be centered.

The Turning Tools

There are two distinct types of cutting tools available for use on the lathe – those which cut on the skew, giving a slicing cut, and those which cut square on, giving a scraping cut. The latter group are often referred to as scrapers, and many people imagine, erroneously, that the scrapings should be dust, not shavings. If we are to cut wood in the way it likes to be cut, with the wood having a polished finish from the tool, then these pure cutting tools must be used.

There are three styles of turning tools which indicate size rather than shape – long and strong, standard and small.

Turning tools need stout handles in either beech or ash, and should have sufficient length and girth to provide strength. At the same time, they must be styled to take the hand comfortably. After gaining experience, many woodturners turn their own handles to suit their particular needs.

The reputable manufacturer forges his blades from high-quality carbon steel, which is skillfully and accurately hardened and tempered to give long edge-holding qualities. There is an increasing tendency amongst some European manufacturers to use highspeed steel, although there seems little advantage in·this.

Chisels for turning have bevels on both sides, and with the cutting edge square or skew, in sizes ranging from ¼–2 inches/6–50mm.

A chisel for parting off, marking out and beading is the parting tool, but a ¼-inch/6-mm square-ground chisel can be used for this work.

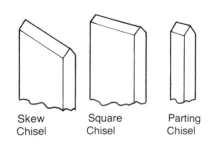

| Skew Chisel | Square Chisel | Parting Chisel |

The turning chisels illustrated, **right**, are, **from left to right**, the diamond, the ½-inch/12-mm skew, three gouges, a parting chisel, a roundnose and a 1-inch/25-mm skew.

The deep long and strong gouges, **above,** are used especially in the making of bowls to remove wood rapidly.

Gouges fall into three groups – spindle gouges, which are rather shallow in section and used for work between centers, deep standard-size gouges for the same work, and deep long and strong gouges.

Shallow gouges, with ends rounded to the shape of the end of the little finger, are used for coving and other spindle work, and also for small bowls and similar face plate work.

An essential gouge for bowl work, is the deep long and strong gouge. This is ground square across, the ideal width being ⅜ inch/9mm. It can also be used in spindle work.

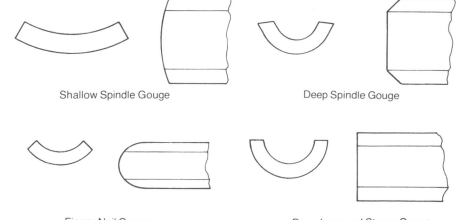

Shallow Spindle Gouge

Deep Spindle Gouge

Finger Nail Gouge

Deep Long and Strong Gouge

Use of Spindle Turning Tools

The shallow gouge is used for roughing the wood down to near size, but to obtain perfect sizing and finishing, the skew chisel is used.

The basic rule of turning is that the bevel rubs the wood, and the cut is obtained by gradually raising the handle to bring the edge of the tool into the cutting position.

The tool slopes into the direction of movement, the center of the chisel or gouge is used and the tool is safely positioned resting on a two-point support (the tool rest and the wood itself). This will always result in completely safe cutting.

All these tools must be sharpened on the ground bevel, which, if it has a high degree of polish, will act as a burnisher and will polish the wood.

The chisel is also used to cut beads, tapers, vees and shoulders. It can also be used for rounding over. The long corner is used to square the ends of spindle work as well as shoulders. The ideal size of a beading gouge is ¼ inch/6mm ground square across.

The round-nosed gouge is ideal for coving and beading.

The illustrations, **right,** show turning tools in use. **1** A shallow gouge is used for preliminary shaping. **2** After the rough sizing smooth work is carried out using a skew chisel. Using a straight chisel, the work is squared off, **3**; rounded off, **4**; and beads cut, **5. 6** A parting tool is used to make a groove. **7** A gouge is used for coving.

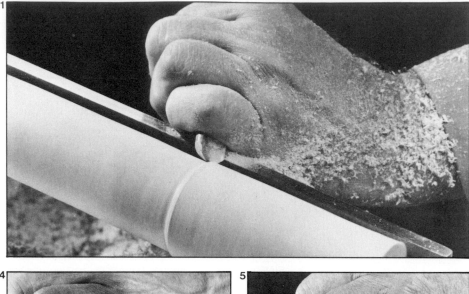

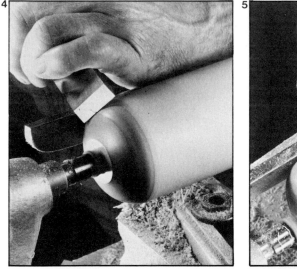

Face Plate Tools

Spindle gouges can be used on bowl work, but the ends must be rounded to the shape of the little finger.

The ideal bowl gouge is the deep long and strong gouge, which is ground square across.

This is sufficiently deep to be used in the tightest of curves without there being any chance of the corners of the tool touching and damaging the turned work.

The tight inside curve of the gouge ensures that the shavings are rapidly removed.

In face plate work, as shown in **8–10,** the inside and outside of a bowl is shaped with round-nosed gouges.

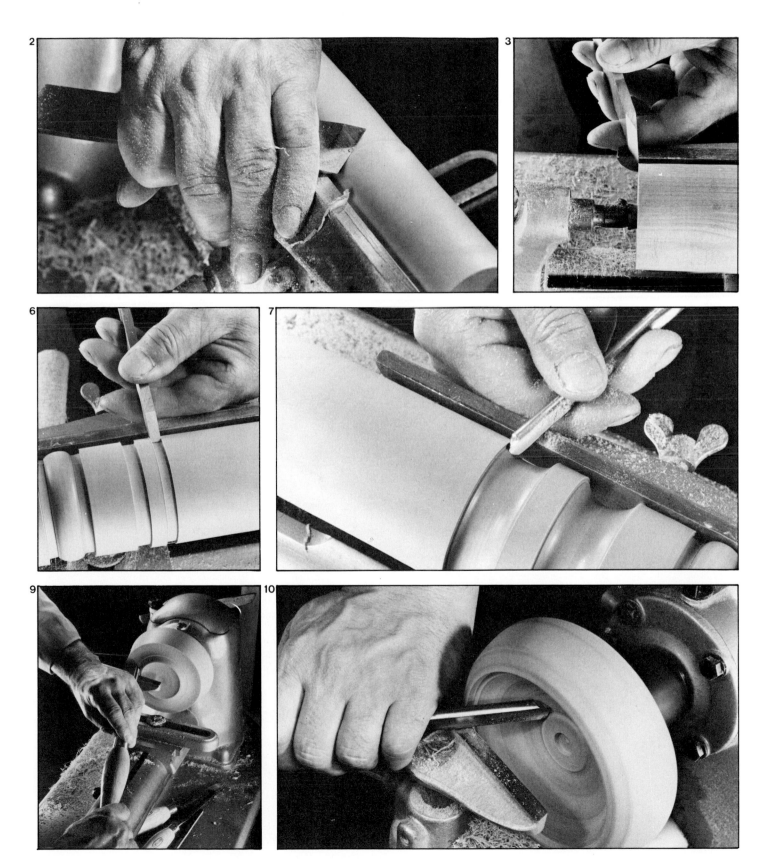

Scraper Type Chisels

These chisels are laid flat on the tool rest with the handle raised to trail the tool. This allows the hooked edge on the chisel to cut the revolving wood. The bevel cannot rub and the resulting finish is inferior to that produced with a correctly-used skew chisel. Many turners re-grind these tools to obtain shapes more suited to their own needs. Many chisels are, in fact, form tools. The correct sharpening of these tools is just as necessary as with the skew chisels and gouges, although the actual cutting edge is produced in a completely different way.

The scraping tools, **below,** have a long and strong heavy section which reduces chattering and flexing. The scrapers, **facing page,** are extra heavy. The weight damps down lathe and wood vibration and so aids fine control.

Scraping chisels are available in a number of shapes, the round nose and diamond point being most common. They have a shorter ground bevel than the normal turning chisel.

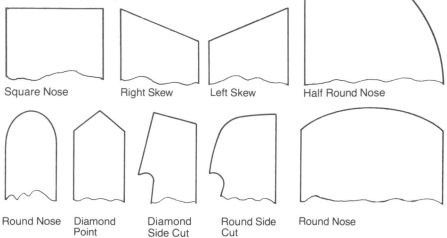

| Square Nose | Right Skew | Left Skew | Half Round Nose |

| Round Nose | Diamond Point | Diamond Side Cut | Round Side Cut | Round Nose |

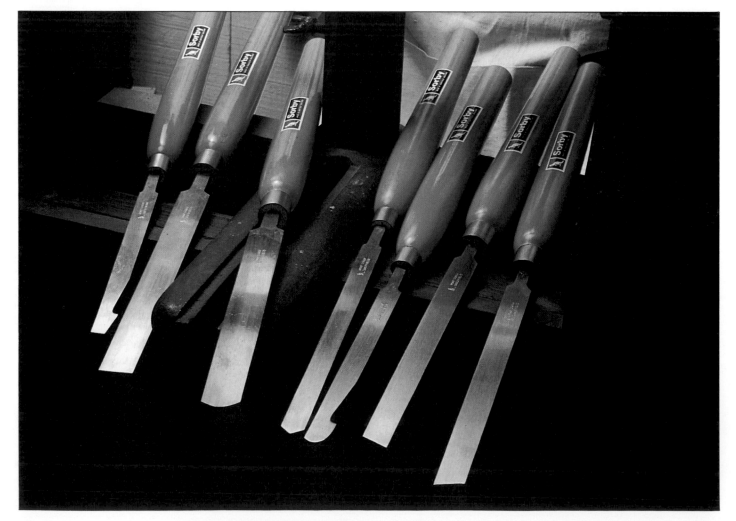

Boring on the Lathe

Various boring tools are necessary for wood turning.

If table and standard lamps are to be bored, a lamp standard shell auger will be needed. They are usually 30 inches/750mm long and ¼ inch/6mm, 5/16 inch/8mm, ⅜ inch/10mm and 7/16 inch/11mm in diameter. This auger cuts easily into end grain, its center lip ensures the bit is central, and if the auger is withdrawn steadily to remove the waste, there is little possibility of it running off-center.

Ideally, large holes for boxes and vases are cut using the saw tooth machine cutter (multi-spur bit in the USA). This is a perfect tool for cutting end or side grain, in any type or condition of timber.

For boring pepper and salt mills, and holes up to 1½ inch/37mm diameter, the flatbit is excellent. It must not be fed into end grain too quickly, otherwise it will run off-true.

Holes of a smaller diameter can be bored using two other types of bit – the wood drill and the lip and spur drill.

The wood drill is similar to the engineer's twist drill, but with a sharper nose and greater chip clearance.

The lip and spur drill has chip clearance similar in design to the wood drill but has a brad point, two spurs to cut the periphery of the hole and cutters to cut and lift the chips.

An auger like the lamp standard shell auger, **above,** will make long holes for electric flex. Unless the tail stock is designed to accept an auger, a long hole-boring attachment like the one, **right,** should be used. This will support the wood as it revolves and at the same time allow the auger to pass through. When boring large holes a tail stock chuck is used. It is fitted with a morse taper and its capacity must be at least ½ inch/12mm to accommodate saw-tooth machine bits like one **below** or similar machine bits such as the flatbit, **below right.**

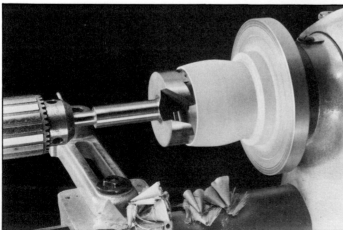

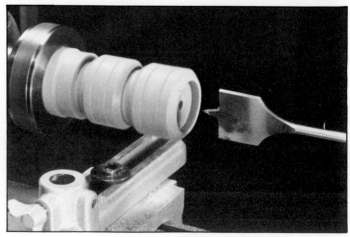

Sizing and Measuring Tools

Another useful tool for the woodturner is the sizing tool. This is used in conjunction with the parting tool or the beading chisel, and can be set to any size. The hooked end is placed over the spindle work and the tool pulled toward the craftsman so that the end of the hook bears against the revolving wood. The edge of the tool is brought into the cutting position in a downward movement.

An essential marking tool is a pair of calipers. One able to take both inside and outside measurements, like the spring-divided calipers, is ideal. A pair of dividers and a rule will also be needed.

To use the sizing tool, **1** set the tool for repetitive work. **2** The tool is positioned with the arm placed on top of the work. **3** The handle of the parting tool is raised until the blade begins to cut the wood. **4** A series of cuts are made for larger work.

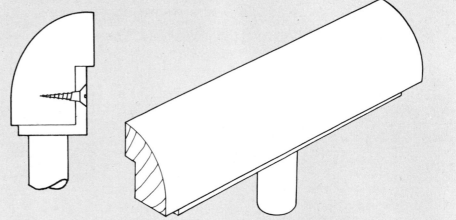

Tool Rests

One piece of equipment supplied with the lathe which has never been perfected by any manufacturer is the tool rest, or tee rest, as it is often called. The shapes are often diverse and fail to give a long straight run to the tool, which is particularly important when chiseling.

To rectify this, the craftsman may care to make his own tool rest, using an angle iron, onto which a peg has been welded, and inserting a length of long straight beech or similar hardwood shaped to fit. A number of lengths of wood can be made up to suit most requirements.

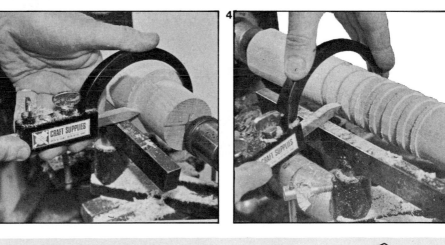

CHAPTER TEN

Carving Tools

*Carving is not only
a very serious and skilled craft, it has
also been widely taken up
as a hobby.*

C ARVING IS NOT ONLY A VERY SERIOUS and skilled craft to which many men have devoted their lives, but has also been widely taken up as a hobby. The encouragement and backing given to this craft in the United States and in Great Britain by manufacturers and suppliers has led to a world-wide recognition of the need for leisure activity in a world of increasing stress.

Walk through the warehouse of any factory which manufactures carving tools, and you would be excused if you fled from the place in blind panic at the sight of so many different shapes, sizes and styles of tools. You might even decide to forget about the craft! Apart from the daunting range of tools, many people feel they are not artistic enough to be able to carve wood. However, carving and sculpting cover such a wide field that even the most inexperienced worker can find an area in which to gain success and satisfaction. The working area need not be a problem either. In Bavaria, the most exquisite work is carried out in tiny backstreet workshops.

Carving tools are the most attractive of the edge tools. In their many shapes and forms, they suggest curves and images and seem to inspire creative activity.

Buy the most basic tools first, and add to them only when necessary. An experienced working carver will own rarely more than 50 tools, with only half that number in constant use. He will have no desire to acquire the 3,000 different carving tools manufactured by one leading company. When setting up, decide before buying whether you wish to carve in low relief, sculpt, chip-carve or whittle, so you do not buy any unnecessary tools.

Of all the craftsmen, the woodcarver has the greatest variety of tools and handles. The European prefers to buy his handles and blades separately, choosing handles to suit his hand and the job. The favorite handle, however, is octagonal and without a ferrule, but in recent years, a couple of manufacturers have produced these handles with an internal ferrule, using a number of hardwoods. In the United Kingdom, the

Carving Tool Shapes

Chisels and gouges are identified by numbers; straight and skew chisels are numbered 1 and 2 respectively, while gouges range from 3 to 12 according to the curvature of the blade.

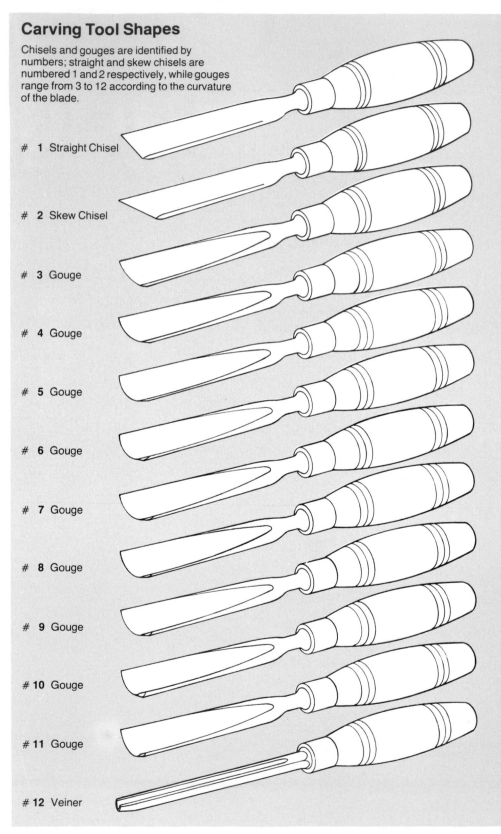

1 Straight Chisel

2 Skew Chisel

3 Gouge

4 Gouge

5 Gouge

6 Gouge

7 Gouge

8 Gouge

9 Gouge

10 Gouge

11 Gouge

12 Veiner

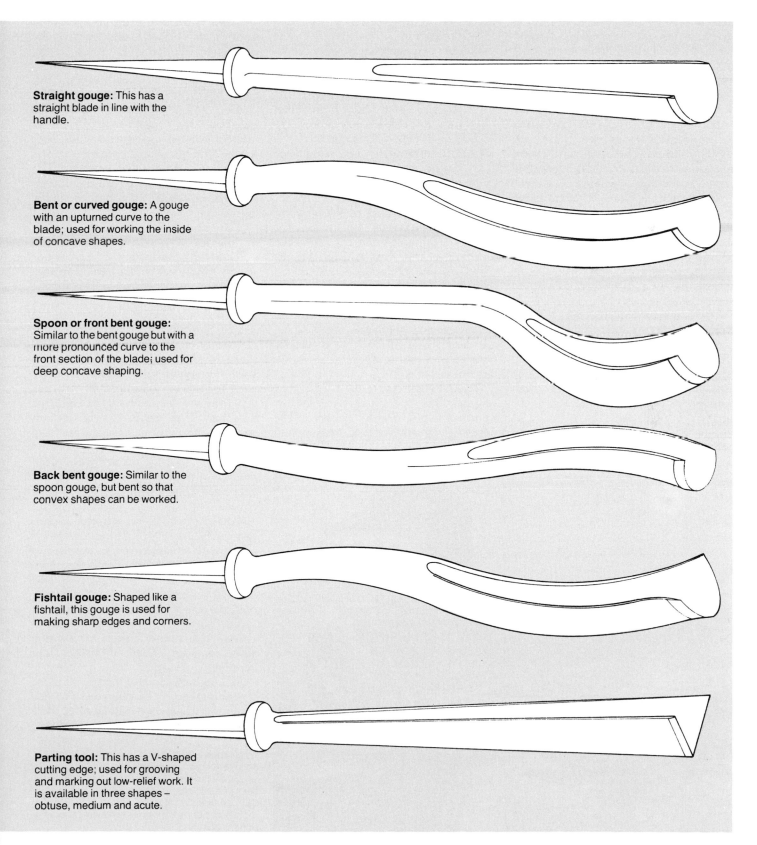

Straight gouge: This has a straight blade in line with the handle.

Bent or curved gouge: A gouge with an upturned curve to the blade; used for working the inside of concave shapes.

Spoon or front bent gouge: Similar to the bent gouge but with a more pronounced curve to the front section of the blade; used for deep concave shaping.

Back bent gouge: Similar to the spoon gouge, but bent so that convex shapes can be worked.

Fishtail gouge: Shaped like a fishtail, this gouge is used for making sharp edges and corners.

Parting tool: This has a V-shaped cutting edge; used for grooving and marking out low-relief work. It is available in three shapes – obtuse, medium and acute.

round handle, made of beech, box-wood, rosewood or jarrah predominates. In the United States both styles are in wide use and are accepted by professionals and amateurs.

Blades vary in finish from the original black and straw to the modern overall bright finish. European manufacturers are introducing ranges which are very highly polished. These might attract the newcomer to the craft, but the dedicated carver would think them unnecessary.

The better the finish, the less likely the tools are to rust, but the essential feature in these tools should be extreme accuracy of shape and first-class finish of the cannel to produce the perfect edge. One manufacturer makes tools almost exactly as they were a hundred years ago.

Looking at a catalog of carving tools, it will become apparent immediately that their reference numbers are progressive and relate only to the shape and cross-section. The early Sheffield List quoted carving tools under such numbers as 3701, 3702 and so on, the

latter two figures indicating the shape and cross section.

The last two numbers are still used by reputable manufacturers and make recognition easy. For example, 01 is No 1 – a chisel ground square across, and 02 is No 2 – a chisel ground on the skew. They were available a hundred years ago in sizes ½–2 inches/12–50mm.

Chisels are available in a number of sizes, both skew and square across, and the ground bevels on both sides are rolled, not flat like the joiner's chisel. Bent chisels are square, or sharpened in the left or right corner, and are used for cutting in difficult situations and in the working of fine detail. The dog-leg chisel is useful in recessed work and the fishtail spade chisel cuts into tight recessed work in low relief.

Special chisels which are seen rarely in today's catalogs are the allongees, which are large taper-blade chisels greatly favored by wood sculptors working on large pieces of wood.

Generally, gouges are classified as slow or quick. The slow gouges are in the low numbers and have fairly flat curves. The quick gouges have deep curves and appear in the higher numbers. They are

available straight or bent, both with the same blade curvature and number, irrespective of their size. Some of the gouges have fantail-shaped blades to give clearance in deep cutting, but generally the blades are straight.

Also, gouges can be of spoon bit or back bent type, again with the curves related. These gouges are extremely useful for difficult shaping where access is impossible with the straight gouge. The back bent gouge will cover difficult convex shapes, whilst the spoon caters for concaves.

The beginner will need a small number of slow and quick gouges in several widths but will not need bent or spoon type tools while he is learning. Sets of smaller tools are available for the beginner, in a restricted variety of shapes.

Vee parting tools can be straight, curved or spoon-shaped, with up to three different included angles, and in a wide variety of sizes. These tools are used for marking out low-relief work, texturing work of all kinds and also for lettering. They are extremely difficult tools to sharpen but beautiful tools to use.

Another tool for texturing and deep narrow cutting is the veiner. This is a very narrow tool with a deep U-shaped curve.

The macaroni chisel is a very useful, but rarely used, tool. It has a wide U-shape with square corners, designed to finish the sides of square recesses, and it is also an excellent bottoming tool. The curved version is useful for working concave recesses.

The fluteroni is a wide U-shaped chisel, with round corners for shaping rounded sides. There is also a bent version.

Where large three-dimensional work is undertaken, it is often best to use sculpting tools. These are stouter tools in larger sizes. Standard gouges with the

Below, a set of six small carving tools comprising straight, skew and bent chisels, straight and bent gouges and a parting tool.

Opposite: 1 An eighteen-tool set of full-size chisels and gouges comprising straight, skew and spoon chisels, ten straight gouges, three bent gouges, one spoon gouge and a parting tool. **2** A selection of block cutters – small carving tools with stubby handles which easily fit into the palm of the hand.

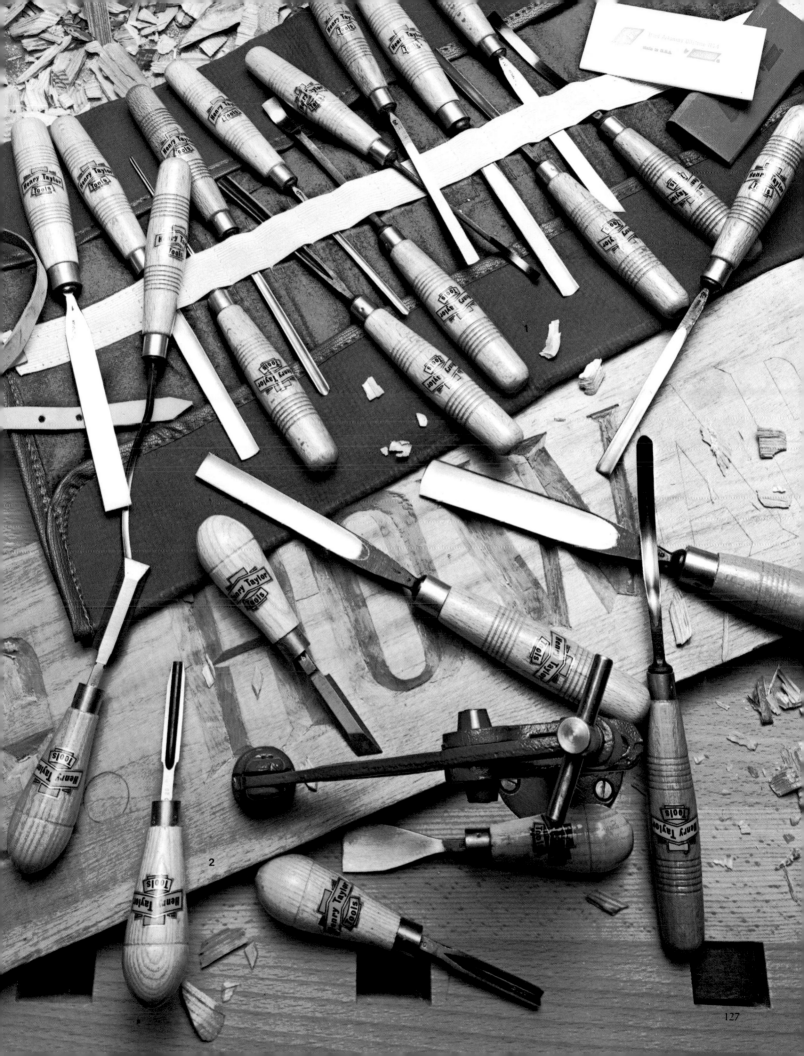

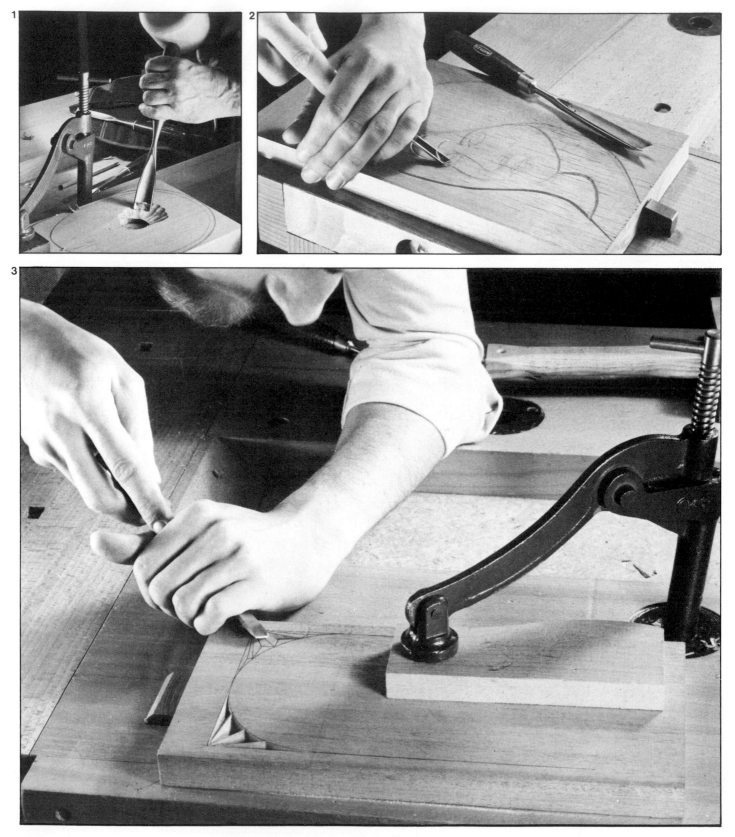

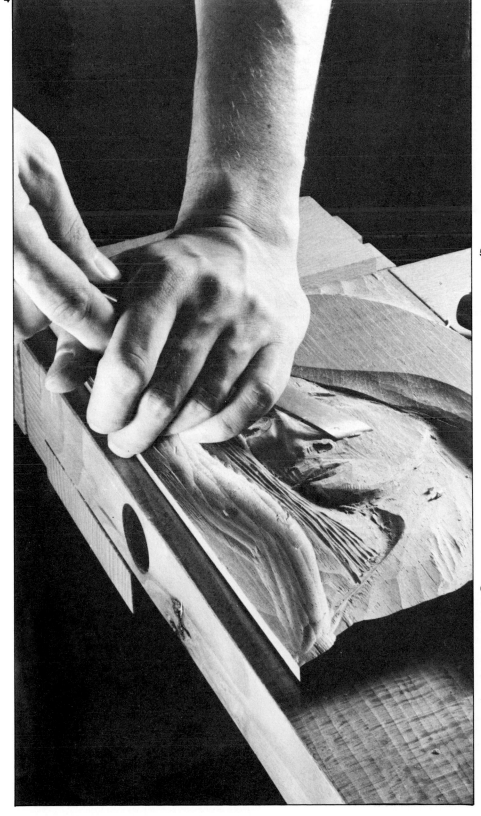

Some carving tools being used in a variety of situations. As a simple guideline, the general form should always be completed before the detail is worked. **1** A straight gouge is used, in conjunction with a carver's mallet, for the initial roughing-out of the workpiece. The tool is first held vertically and is then quickly lowered into a more horizontal position. The workpiece is held by a bench holdfast. Mallets are used for roughing down, and the finishing is done by hand. **2** Cutting a groove with a vee parting tool. The wood is being secured by bench dogs. **3** A straight chisel is used to cut a deep angular design in a workpiece which is being held by a bench holdfast. **4** A smooth surface is obtained with a skew chisel. This tool helps to give a slicing cut. **5** A concave shape is hollowed out with a bent gouge – the only tool able to do this – while being held in an end vice. **6** Using a 'quick' gouge to cut a scalloped shape.

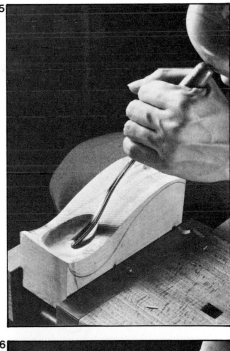

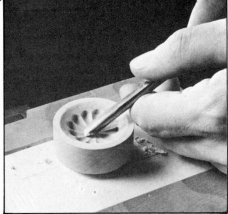

normal tanged blade and wooden handle are popular with sculptors. One range has long blades and standard-length handles, often with leather caps. They will stand up to the very heaviest work.

Many carvers, particularly in North America, use knives only, with great skill. This is called whittling – a term which suggests that little skill and attention to detail is needed. However, this impression is quite wrong, and can be proved so by inspecting whittled wood. The knives used are of high-quality steel, and are available in a variety of blade shapes to meet the needs of the whittler.

The carving of domestic ware such as dishes, trays and bowls can usually be done with the normal carving tools. The bowls of small spoons and suchlike can be cut with the carver's hook.

In the past, the hook had a long handle which could be tucked under the arm. The modern version is adequate for small work and can be used for scooping out curved recesses. A similar tool is the scorp, which has a circular blade.

A collection of beautiful tools which became available recently are carving planes, 1⅛ inches/28mm long and 1 inch/25mm wide. With polished metal bodies and integral handles which extend comfortably into the palm of the hand, these planes are available in three sole styles – flat, lengthwise curve and spoon. When correctly sharpened and finely set, they are capable of the finest finishing work.

Many carvers and wood sculptors working on large forms need a carver's adz for the speedy roughing-down and removal of waste. This is

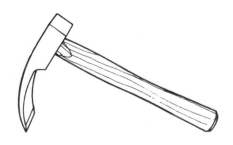

Top, a twelve-tool set of light carving tools. These are the same quality as the full-size tools but are smaller and lighter. The mallet has a *Lignum vitae* head and an ash handle. **Above,** carving planes are available with flat, curved and spoon soles and are used for fine finishing work. **Right,** a quick way to rough down and remove waste is to use a flatbit attached to a power drill. **Far right**, the carver's adz is a more traditional tool for quickly removing waste. This adz has a small blade, 2 inches/50mm wide and an 8½-inch/212-mm long ash handle.

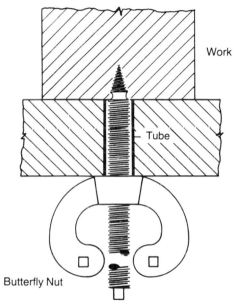

Woodcarving work must be well-secured. This woodworker's bench, **left,** allows work to be held by dogs or stops inserted in slots in the bench top and the end vice. **Below and bottom,** the woodcarver's screw is used to secure work in an upright position where the carver can move around it. The cross-section shows how the screw secures the work to the bench top and where a tube can be inserted to prevent wear.

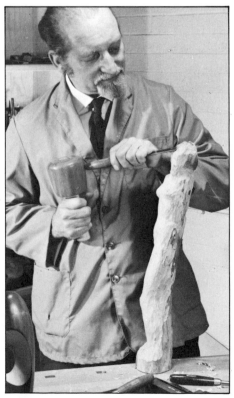

available with both straight and curved edges, and is usually double-ended. The handles are made of ash or hickory and are at least 9 inches/225mm long.

The woodcarver's hatchet is used when sizing down blocks, as it removes waste quickly.

Rifflers are very useful shaping tools used by all woodworkers, but especially sculptors. They are available in two styles – double-ended, with rasp teeth at one end and file teeth at the other, or with one blade of either rasp or file teeth, fitted into a wooden handle. The blades are about 2 inches/50mm long, with tooth sizes in second cut and bastard.

These tools are ideal for finishing curved surfaces and fine work generally.

The woodcarver must be able to secure his work safely when carving and, if working in three dimensions, needs to move his work around constantly.

In the home workshop, a block of wood can be fixed to a bench, using a woodcarver's screw. This is screwed into the underside of the block, the screw passed through a hole in the bench and secured with a butterfly nut. A square hole in its wing fits a square boss on the end of the screw and helps turn the screw into the block.

Using this tool, the sculptor can work around the wood without obstruction.

If the screw is used frequently, it will save wear and tear on the bench top if a piece of tube of appropriate size is inserted, as shown in the diagram.

When preparing the hole in the block to receive the woodcarver's screw, first bore a hole the diameter of the screw to a depth of at least 1 inch/25mm. Then, bore a hole half the diameter, into which the screw will be turned. The first hole will help to give extra support.

Small work can be held down on a block secured in the vice or on the bench top using a carver's clip. This is a simple turn button with one serrated edge, the other bent and serrated with a center screw to turn into the bench.

Carvers seem to have been neglected by vice-makers. The traditional vice is

Making a Scopas Chops

The fixed jaw is housed and glued into the base and reinforced with two buttresses. The sliding jaw fits in a channel produced by gluing two strips of wood to the base; steps are cut from each of the bottom corners of the jaw to allow a 'sliding fit' and buttresses are attached to the back edge. Two guides are housed into the fixed jaw and the tail piece for extra support; the guide housings in the sliding jaw must allow a good 'sliding fit'. The tail piece is secured to the base by glue and screws. A 16-×1-inch/400-×25-mm bench screw passes through 1-inch/28-mm holes in the sliding jaw and tail piece and the end is located by a shallow 1-inch/25-mm hole bored into the inside face of the fixed jaw. To position the nut in the sliding jaw, the bench screw is located in the fixed jaw and, by a split collar, in the tail piece. The position of the nut can then be marked on the back of the sliding jaw. The vice is secured by a bolt passing through the bench top and held by a butterfly nut. Use hardwood and seal with a polyurethane varnish; rubber strips attached to the jaws will protect both the vice and the work.

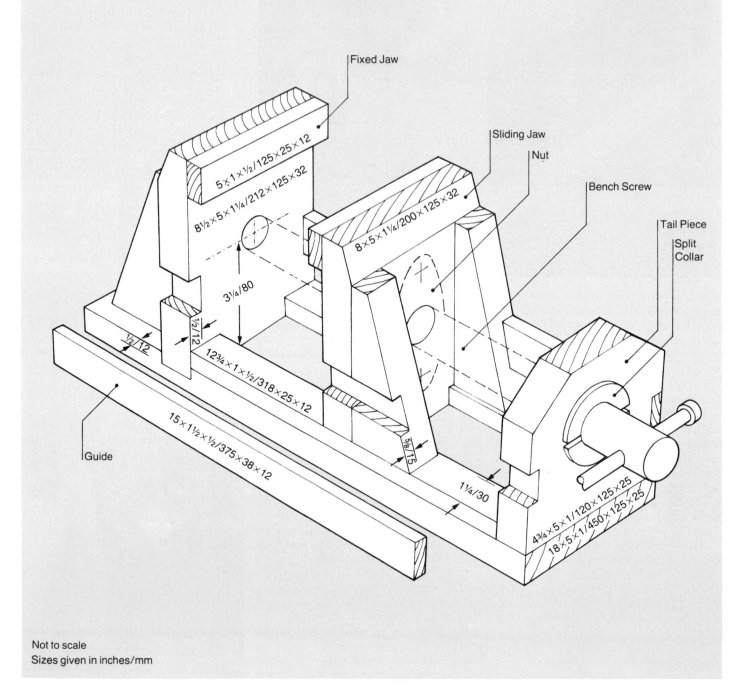

Fixed Jaw

Sliding Jaw

Nut

Bench Screw

Tail Piece

Split Collar

Guide

5×1×1½/125×25×12

8½×5×1¼/212×125×32

8×5×1¼/200×125×32

3¼/80

½/12

½/12

12¾×1×1½/318×25×12

15×1½×½/375×38×12

⅝/15

1¼/30

4¾×5×1/120×125×25

18×5×1/450×125×25

Not to scale

Sizes given in inches/mm

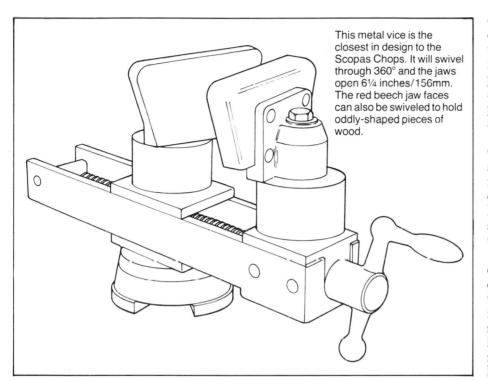

This metal vice is the closest in design to the Scopas Chops. It will swivel through 360° and the jaws open 6¼ inches/156mm. The red beech jaw faces can also be swiveled to hold oddly-shaped pieces of wood.

often quite inadequate, particularly in depth. A vice which was designed for the woodcarver is the Scopas Chops. It is made of wood, with deep jaws and protective leather strips, and has a steel screw and nut and another screw which passes through a hole in the bench top to secure the vice.

A metal vice similar to the Scopas Chops might be considered by the serious carver. This has deep wooden-faced jaws which slide along a solid channel bed. It can be fixed to the bench like the Scopas Chops. The vice can be swiveled through 360° and is probably the nearest equivalent to the Chops.

The woodcarver's kit would be in complete without a mallet. It is vital to choose one that is well-balanced and of the correct weight, if it is to be used to its best advantage. The European beech mallet, with its beech head and ash handle, weighs about 1 lb/0.4kg, and most people find this adequate. Its handle is designed to fit the hand and its curved face counters indentation and avoids surface splitting. The pliant ash handle resists shock.

Many carvers prefer the harder and often heavier *Lignum vitae* head. Often the mallet is made with an integral handle, which some people find shocks the hand.

Lignum vitae resists indentation and will give life-long service if it is looked after properly. When not in use, store the mallet in a tight plastic bag to prevent the wood drying and cracking.

The round head of the carver's mallet is used in a series of short blows from the wrist, quite different from the swing given to the joiner's mallet. The rounded shape ensures that any point of the mallet which strikes the tool will perform adequately. The carver can thus ignore the mallet and concentrate on the cut being made. He need not at any time refer to the position of the head as he would need to do were he using the flat-faced joiner's mallet.

When buying these tools, choose them carefully, and only buy those which are absolutely necessary.

Left, the holdfast will grip large workpieces on the bench; the shaft is held by a metal collar housed in the bench top.

Veneering Tools

*Nearly all the trees
giving beautifully colored
and grained timber
can be obtained in veneer
form.*

V

ENEERING COULD BE DESCRIBED as
covering over a poor- or inferior-
grained timber with a thin layer of more
decorative wood. However, this is not
strictly true when describing some of
the very beautiful veneered work of the
past, where skilled constructional work
demanded high-quality timber like
mahogany.

Nearly all the trees giving beautifully-
colored and grained timber can be
obtained in veneer form. Many timbers
from the smaller trees would not otherwise be seen.

Veneers are made by either peeling or
slicing the log, so the majority of the
timber log is used. The slicing method
produces 'figure' which, in timbers like
oak, is described as silvergrain.

In the past, veneering was shrouded
in a veil of mystery, and the necessary
tools were not readily available. Today,
many man-made boards are veneered,
using the traditional methods, and are
used in the mass production of furniture. Man-made boards, like multi-ply
woods, make excellent base materials
for the home craftsman to veneer.

Inlays and Marquetry

Banding is another form of veneering,
where many tiny pieces of veneer are
placed together to form a pattern. The
machine method of making banding is
to glue long strips of timber edge to edge
and then to saw the wood across the
grain into thin strips. These long
patterned strips can be used with
veneers or they can be placed in shallow
grooves in solid timber either as a
decorative border or in geometric or
other patterned form.

Veneers are used to make up pictures
in a craft known as marquetry. Here,
both the color and grain give scope to
the artist craftsman. The absence of
colors like bright greens and magentas is
solved by using dyed timbers, sycamore
being a particularly popular timber for
this purpose.

The tools of the craft can be broadly
divided into two groups – those for
cutting and those for laying. The
extreme thinness of veneers means that
many of the traditional timber-cutting
tools cannot be used.

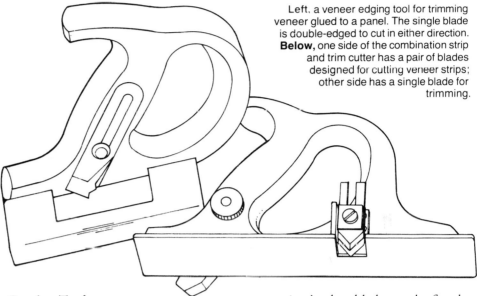

Left, a veneer edging tool for trimming veneer glued to a panel. The single blade is double-edged to cut in either direction. **Below,** one side of the combination strip and trim cutter has a pair of blades designed for cutting veneer strips; other side has a single blade for trimming.

Cutting Tools

The veneer knife is probably the best tool to use for cutting because it will produce a perfect edge, which is particularly vital where one edge butts up to another. The knife must have a curved edge to give a perfect slicing cut without tearing. There are many suitable general-purpose knives on the market. However, the best tool for the serious craftsman is the trimming knife, which is available with several types of blade.

Veneers can also be sawn with a veneer saw, which has a rectangular blade with a round handle attached to one of the short sides. The blade is curved on both long edges, with the teeth on one edge straight-edged, and those on the other, taper cut. The saw is used in conjunction with a straight edge.

The traditional fret saw is used to cut small and intricate shapes. It must have a fairly shallow throat if only tiny work is to be cut, but use a deep-throated saw if cutting well into the veneer. A small, adjustable fret saw is most useful in tight situations and the frame adjustment

1 An adjustable fret saw, also known as a piercing saw. The blade has 32 points per inch/25mm. **2** The traditional veneer saw has a curved blade to prevent the corners digging into the veneer. **3** Veneer punches are used to replace damaged areas of veneer. They have an irregularly-shaped cutter and a spring-loaded plate which removes the veneer. The handle is hooped with a metal striking plate.

means that broken blades can be fitted – very useful as blades break very easily in veneer work.

Use a combination strip and trim cutter to cut strips of veneer. This tool has an integral body and a handle which is shaped like that of the wooden jack plane. It is fitted with a cutter at each side, which can be adjusted using shims. Small clamps hold the blades in place and a screw holds the unit securely. The cutter on one side is used for trimming veneers and the other cuts strips of veneer from 1/12–5/16 inch/2–8mm wide. Always use this tool with a straight edge as a guide.

The best way to trim veneer edges is with an edging tool. It is similar in design to the strip and trim cutter but has only one cutter which has a spear point to cut in both directions.

Small blemishes frequently occur in veneers and spoil the appearance of the finished work if they are not removed first. The craftsman often cuts these out in diamond, circular or rectangular shapes which only emphasize the repair. The best way to avoid this is to use a veneer punch. It has a thin, hardened-steel cutting edge which removes an irregular-shaped piece of veneer. This is replaced with another piece cut from a perfectly matching veneer. The wooden handle can be struck with a hammer, as it has a metal cap. Punches are available in a number of shapes and sizes.

Inlaying Tools

The cutting gauge is the tool most commonly used for cutting grooves for inlays and beading. It has a small cutter, shaped to cut in both directions with a slicing action, and a stock which can be set at any distance along the beam and secured in place with a thumbscrew. The cutting gauge can be converted into a scratch stock, which has a small cutter to cut a narrow groove into which an inlay and banding is glued.

A scratch stock can be made using two pieces of timber, a number of small bolts and a cutter made from an old hacksaw blade.

The inlay cutter is a superb German tool, made of beech and brass, which is used to cut shallow straight or curved grooves to receive inlays. The fence is adjustable up to 6 inches/150mm from the edge of the work with veins between $^1/_{12}$–$^5/_{16}$ inch/2–8mm wide. It has two blades which cut the sides and another which removes the waste between the veins. Shims give minute adjustments. This tool is a perfect example of a tool designed to meet a specific need exactly.

A tool designed for violin-makers, but which is equally suitable for general craftsmen is the purfling cutter. This is capable of cutting very small rebates and grooves.

The correct tool, as sharp as possible, must be used if veneering, inlaying and banding is to be done with complete success. Veneering is an intricate craft, but extremely accurate marking out and cutting will contribute greatly to its final perfection.

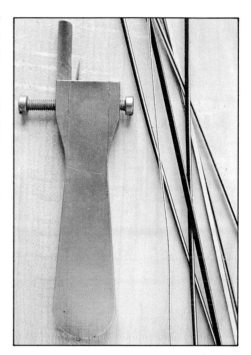

The special inlay cutter, **facing page**, is designed to cut narrow grooves with two cutter blades in conjunction with the clearing-out cutter. The cutter of the purfling tool, **right**, is held and adjusted by two opposing screws. It is used to cut small rebates and grooves on stringed instruments, as shown **below**.

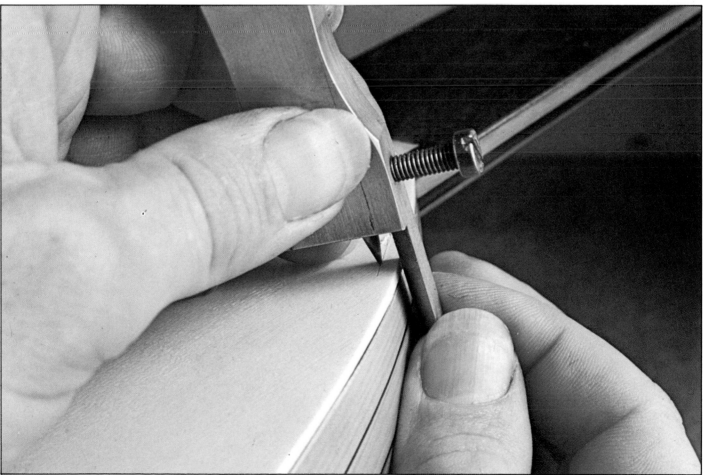

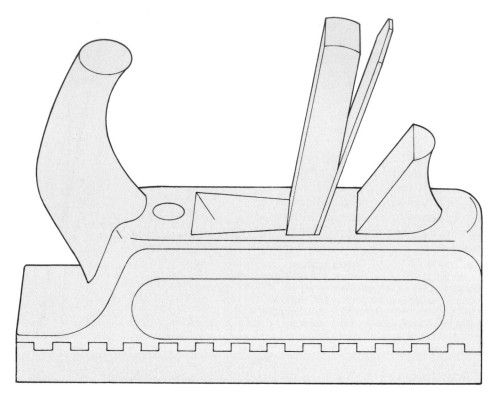

Laying Tools

Veneer is usually laid on a perfectly flat surface but, at the same time, the adhesive must be keyed into the timber. Use a toothing plane for this. It looks like the traditional wooden smoothing plane but its cutter is held almost vertically, and has a serrated edge to score the surface of the wood. Any excess glue or air bubbles can then run into the grooves cut in the wood, allowing the veneer to lie flat.

Alternatively, fit a double-handed scraper with a toothing blade.

Traditionally, veneers were laid using hot fish or animal glue prepared in a double glue pot. The older craftsmen still prefer this type of glue because it gives greater assembly time, and if any glue comes through the veneer, it can be washed off easily without staining or marking. Modern glues need not be

Left, a traditional toothing plane. A toothed cutter held almost vertically scores the surface of the wood, allowing excess glue and air bubbles to escape.

Making a Scratch Stock

A scratch stock is a simple tool for cutting small moldings and inlay grooves. To make one in the workshop, take an L-shaped block of wood about 6 × 2½ × 1 inches/ 150 × 62 × 25mm, and cut it in half through the thickness. The two halves are then held together by screws and the long edge of the notch is slightly rounded. A cutter is made from a strip of steel such as an old saw blade and is filed or ground to the required shape. The edge is ground square and the cutter is held in the tool by tightening the screws. The scratch stock is then simply worked backwards and forwards with a scraping action, keeping the short edge of the notch against the edge of the wood, **below.**

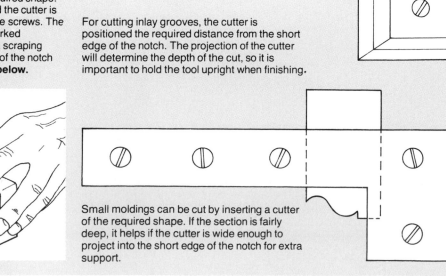

For cutting inlay grooves, the cutter is positioned the required distance from the short edge of the notch. The projection of the cutter will determine the depth of the cut, so it is important to hold the tool upright when finishing.

Small moldings can be cut by inserting a cutter of the required shape. If the section is fairly deep, it helps if the cutter is wide enough to project into the short edge of the notch for extra support.

heated and are extremely efficient, but they can stain the veneer if not used very carefully.

The veneer is pressed into place with a veneering hammer. The traditional hammer has a wooden head and handle with a metal strip inserted along the edge of the head. This smooth strip rubs the surface of the veneer, pushing out air and surplus glue to ensure a perfect union of the veneer and the groundwork. A modern version has a solid metal head.

The roller is an alternative to the hammer. It is similar to the wallpaper-hanging roller, and is particularly useful for laying the edges and jointed edges in veneering.

Whenever possible, use a press to ensure the veneer is laid perfectly. One can be made easily using a vice bench screw or a veneer press screw.

Right, fish or animal glue is bought as a cake or, more usually, as pearls. It is applied hot after heating slowly in a double glue kettle. It should never be allowed to boil.

Methods of Laying Veneer

Using fish or animal glue, the veneer can be pressed down with a veneering hammer, **below.** It is used with a zigzag movement from the middle towards the edges, preferably in the direction of the grain. Surplus glue is pushed out at the edges. Larger pieces of veneer can be pressed down with cauls, **right.** Cauls are strong, flat boards which are laid above and below the veneer and groundwork and pressed together with clamps and cross-bearers. The cross-bearers are tightened in the order indicated to squeeze the glue from the middle towards the ends.

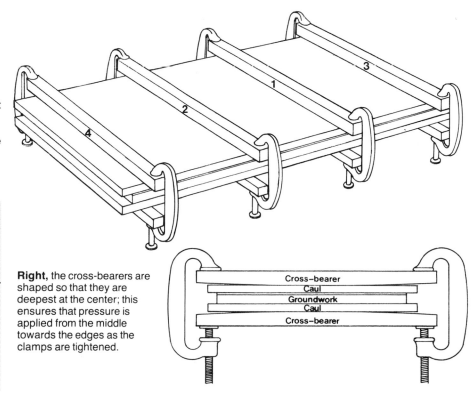

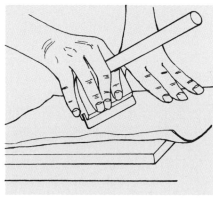

Right, the cross-bearers are shaped so that they are deepest at the center; this ensures that pressure is applied from the middle towards the edges as the clamps are tightened.

| Cross-bearer |
| Caul |
| Groundwork |
| Caul |
| Cross-bearer |

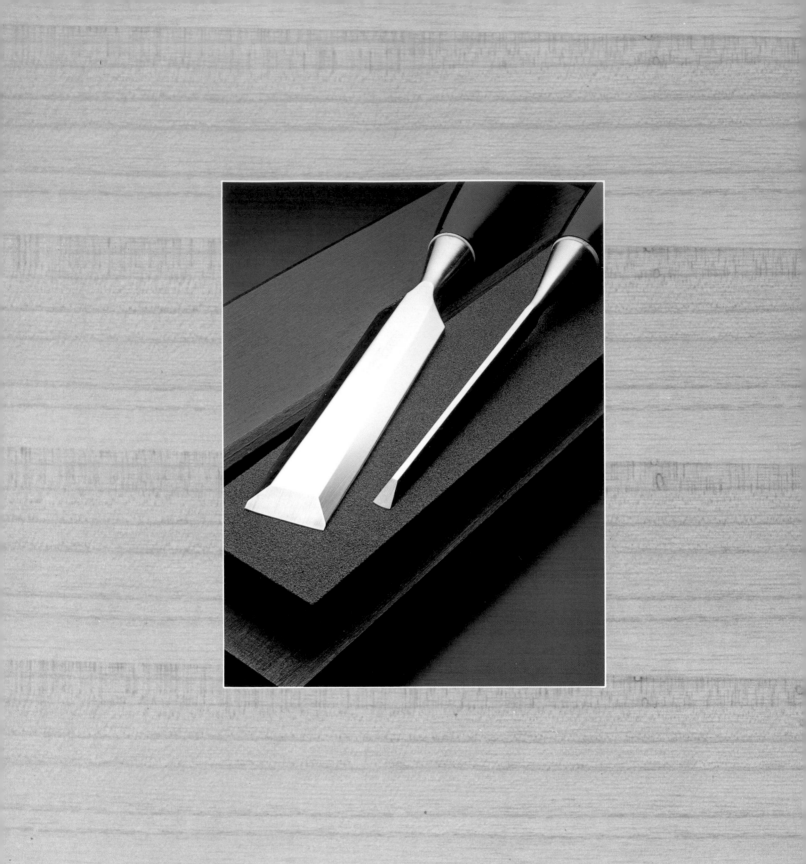

CHAPTER TWELVE

Maintenance

*Perhaps the greatest
deterrent to success with
woodworking
tools is not keeping them
perfectly ground and
sharpened.*

A GREAT DEAL of the frustration and disappointment which people experience with woodcraft tools is caused by lack of guidance and instruction in the choice of those tools in the first place. Perhaps the greatest deterrent to success, however, is the failure to appreciate that, no matter how perfect the tool is, if it has a cutting edge, then it must be perfectly ground and sharpened. Most tools are accurately-ground at the factory but few are ready-sharpened and, anyway, all will need to be sharpened from time to time.

No one seems to like sharpening tools, yet the satisfaction of using a really sharp tool is appreciated by everyone. Cutting should be effortless, and a correctly and perfectly sharpened tool is also much safer.

The equipment for grinding and sharpening can be very simple or quite sophisticated, depending on the space and money available. It is unlikely that you will be able to have your tools re-ground and sharpened by a professional, so you must ensure you have the right equipment to do it yourself.

Using Stones

Bench oilstones are used to sharpen chisels, plane cutters and other straight edges, and slipstones are used for gouges and curved edges.

Stones are divided into two groups – man-made and natural. The best are natural stone quarried in Arkansas in the USA. These are available as bench stones, in 6×2×1-inch/150×50×25-mm or 8×2×1-inch/200×50×25-mm sizes, in four grades. Washita is a fast-cutting grade but the best one for general-purpose is Soft Arkansas. Hard Arkansas is used for finishing a fast edge produced on a Washita or Soft Arkansas stone. Black Hard Arkansas is extremely fine and hard, and is ideal for sharpening surgical instruments.

Man-made stones are in common use and appear under the name India or Crystolon. They are available in coarse, medium and fine grades and in a variety of shapes and sizes.

A recent addition to the range of man-made stones is the Ultimate Diemond. This stone is guaranteed against wear and can be used with either water or oil as a lubricant.

A modern sharpening material is the rubberized sharpening stick and wheel. Here, the cutting grit – silicon carbide – is bonded in neoprene rubber. This produces a very fine cutting edge and, at the same time, a polished bevel ideal for carving and sharpening tools.

Slipstones are available in all the foregoing materials. When you buy one, examine it carefully and choose it according to the tool to be sharpened.

Round-edge slipstones usually have two differing radii on the long edges of the slip, and the flat side is used for normal honing.

All bench stones should be wiped clean and boxed after use, and an oilstone bench should be similarly covered. Make a box to house the slip-stones, with a slot for each one. If these become clogged, boil them in water. Move the tools across the bench stones when sharpening them, to distribute wear, and periodically rub them over a large flat stone to level them out. These stones are expensive and brittle – don't drop them or leave them where they are likely to fall to the floor.

Many people have great difficulty in keeping the right sharpening angle on plane cutters and chisels. A honing guide will correct this fault, by maintaining the angle. One such guide made in Britain is adjustable for blades up to 2¹¹⁄₁₆ inches/65.4mm and for angles of 25–30°.

The majority of man-made stones are already impregnated with oil, but all stones should be used with oil to float away the metal and keep the tool cool. Use a fine, thin, non-drying oil – a number of proprietary brands are available.

A multistone is a good investment. It has a rotating frame fitted with three stones, which are fine, medium and coarse. Each stone can be positioned for use and locked in place, and the two stones not in use lie in a bath of oil. Usually, the manufacturers offer a choice of stones and grades, but all are 11½×2½×½ inch/287×62×12mm in size.

Polish the sharpened tool edges either with rubberized tools or a leather strop. This can be made up with a strip of leather glued to a narrow board, the end of which is shaped to form a handle. These are available ready-made, and the most useful one is four-sided. One side is dressed in emery, the second and third sides are medium- and coarse-dressed leather and the final side is undressed leather. Cutting and polishing compounds can be bought for dressing but various grades of emery can be mixed with tallow or other medium to provide suitable cutting agents.

Sharpening Chisels

Rest the ground face flat on an oilstone which has been lubricated with light machine oil to keep the tool cool and float away metal particles. Raise the chisel to form an angle of 30° with the stone. Move the chisel backwards and forwards in a figure of eight movement along the full length of the stone, until a burr forms on the flat side of the chisel and extends along the full width of the cutting edge.

Place the chisel flat on the oilstone, ground face up, and lightly stroke it up and down to remove the burr.

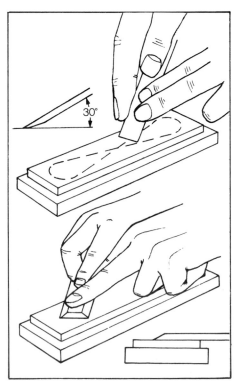

Sharpening Plane Cutters

Many craftsmen prefer to sharpen on the 25° ground bevel but usually a 30° angle is formed when sharpening.

Rest the ground bevel flat on the oilstone and raise the cutter to form a honing angle of 30°. Move the cutter backwards and forwards along the full length of the stone until a burr is formed on the flat side. Place the cutter flat on the oilstone with the bevel uppermost. A few light strokes along the stone will remove the burr. The cutter is now ready for use.

If sharpening a smoothing plane cutter, slightly round each edge, and if sharpening a jack plane cutter, make sure the cutting edge has a slightly convex shape.

Sharpening Gouges

Large gouges are best sharpened using a gouge slipstone. These are tapered stones and have one convex and one concave side, to hone both in- and out-cannel gouges.

The out-cannel gouge is sharpened on a sharpening stone. Place the ground bevel on the stone and rub it backwards and forwards along the length of the stone, whilst slowly rotating it from edge to edge at the same time, to produce a burr along the entire cutting edge.

Remove this burr with a slipstone, keeping it flat against the inside of the gouge. Always keep the edge square at the ends, so that when pressed onto a piece of wood, the gouge leaves a clear imprint on its surface.

Finish the bevel finally by rotating the gouge at a 90° angle along the stone.

To sharpen an in-cannel gouge, choose a round-edged slipstone of the same radius as the gouge. Carefully fix the lightly-oiled stone in a woodworking vice or a slotted block. Rest the ground bevel of the gouge flat on the curved edge of the stone. Raise the gouge 5° and rub it backwards and forwards along the stone until a burr appears along the full length of the cutting edge.

Oil a flat oilstone, and rotate the gouge flat along it to remove the burr. The gouge is now ready for use.

Use the full length of the stone when honing a plane cutter.

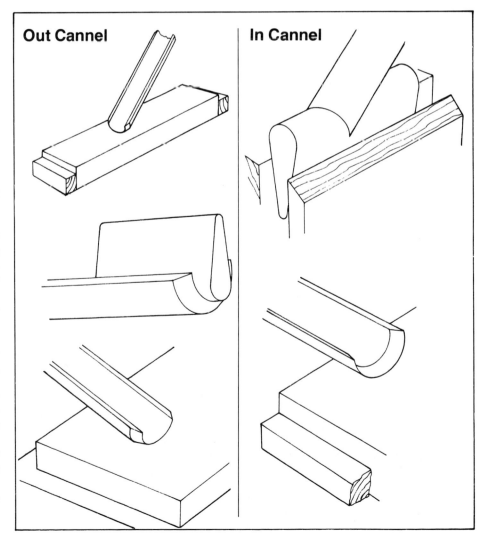

Out Cannel **In Cannel**

145

A selection of stones, both natural and man-made, which can be used to sharpen tools.
1 Black Hard Arkansas – a slow-cutting stone which produces a razor-sharp edge. **2** Hard Arkansas – the best all-round stone for the final polishing of a sharp edge. **3** The best general-purpose stone – Soft Arkansas. **4** Washita – the most rapid cutter of the four grades of natural stone. **5** A combination stone of Hard Arkansas and Washita. **6** A stone made of combined man-made coarse and fine India, which is roughly equivalent to Washita. **7** Another man-made stone – the faster-cutting Crystolon. **8** A honing guide ensures accurate bevels of 25° and 30° when sharpening. It can take blades up to 2^{11}/$_{16}$ inches/66.5mm wide.

If a slipstone is too sharp when new, rub a lead pencil into it to take off the edge of the grit.
1 Four slipstones – Hard Arkansas, Soft Arkansas, Washita and Crystolon. **2** A Hard Arkansas grooved stone, very useful for honing small, steep gouges. **3** A Hard Arkansas knife file and a Hard Arkansas triangular file. **4** A medium India knife file and a fine India auger stone. **5** A set of four Hard Arkansas stones in a variety of shapes. **6** Hard Arkansas file set with square, knife, round, diamond and triangular stones. **7** A fine India round-edge cone gouge slipstone. **8** An ax sharpening stone of Hard Crystolon on one side and Medium Crystolon on the other.

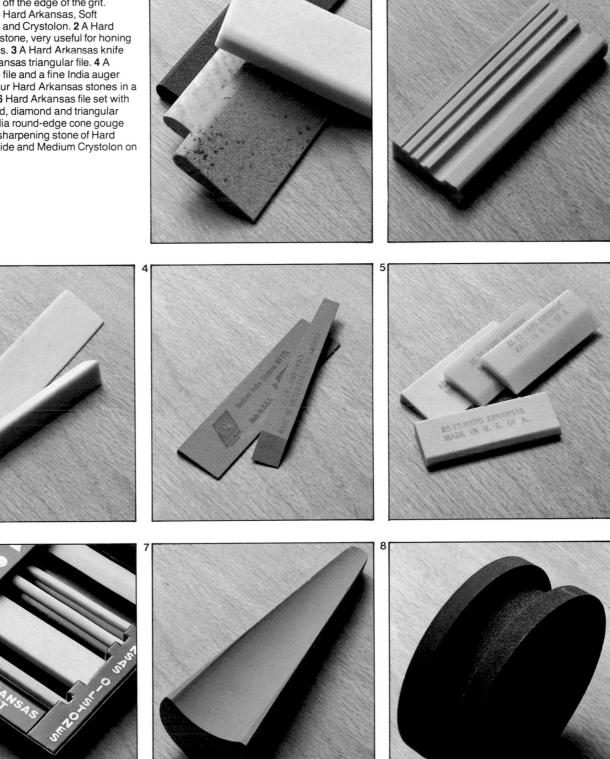

Sharpening Turning Tools

Turning chisels are ground on both sides and have an included angle of 30°. Place the ground bevel flat on a normal stone and move it backwards and forwards along the stone, preferably in a figure of eight movement to distribute wear over the whole surface of the oilstone. Do this on both bevels until a burr appears along the entire length of the cutting edge and finally breaks away. The tool will then be ready for use. The parting tool is sharpened in the same way.

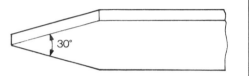

Turning gouges are sharpened in the same way as out-cannel gouges. The tool is sharpened on the ground bevel and then buffed to obtain the best-possible finish.

Scraping chisels are ground at a 10–15° angle on one side only. The top side of the chisel should be absolutely flat, and the bevel smooth. Rub it with an oilstone to produce this if necessary.

The scraper chisel is not sharpened like normal edge tools, as it is first burnished, and then a burr is produced to form the cutting edge. A small piece of hard round steel fitted in a file or chisel handle can be used as a burnisher or ticketer, but a 4-inch/100-mm long file with no teeth and rounded corners makes a much better tool.

Place the chisel flat on the bench, bevel side down, and draw the burnisher backwards and forwards along the edge, pressing quite hard but making sure all the time that the burnisher lies dead flat on the blade. Do this several dozen times to consolidate the metal, and give a tougher surface.

Now set the chisel in a vice, with the ground edge facing you. With the burnisher lying flat on the ground bevel, burnish backwards and forwards, again pressing quite hard. As the burnishing proceeds, raise the handle slowly until the burnisher lies almost horizontal. This sets up a burr and makes the tool ready for use.

Sharpening a Scraper Chisel

The sharpening of a scraper chisel is illustrated **below**. **1** Scraper chisels are ground at a 10–15° angle on one side only. **2** The cutting edge is burnished with a ticketer. **3** The chisel is placed flat on the bench, bevel side down, and the metal consolidated with the ticketer. **4** It is then set in a vice and the ground edge burnished at an angle of 10°. **5** Gradually raise the burnisher until it lies horizontal.

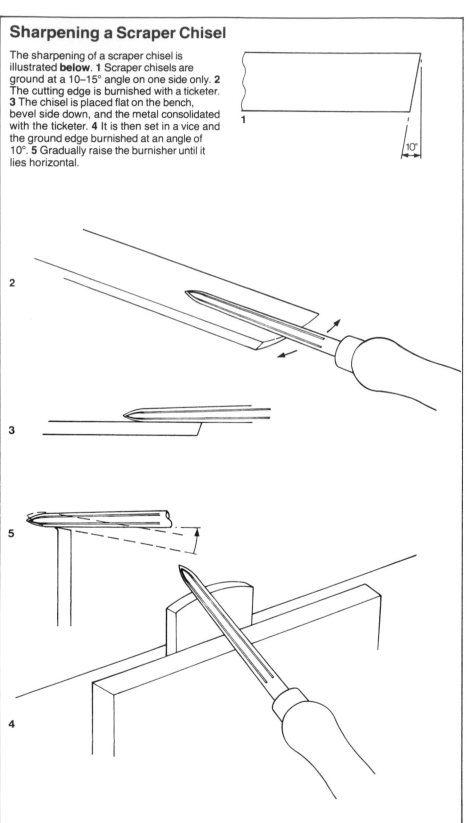

Sharpening Carving Tools

These tools can be divided into three groups – gouges, chisels and vee tools. They can be compared very closely with the woodturning tools, where the tool not only cuts the timber but also burnishes it at the same time. Ideally then, the tools should be sharpened on the ground bevel and buffed to arrive at the best-possible finish.

The gouge is sharpened in exactly the same way as the firmer gouge.

A final finish is given to the bevel by rotating the gouge, holding it at right angles to the stone and moving it lengthwise along the stone.

The chisel has a rolled or curved cannel (bevel) unlike the chisel of the joiner or cabinet-maker. This can be maintained during sharpening or a flat bevel formed.

Place the ground bevel on the stone and move it backwards and forwards along the stone, preferably in a figure of eight movement to distribute wear over the whole surface of the stone. Do this on both bevels until a burr appears and finally breaks off, when the tool will be ready for use.

To sharpen the vee parting tool, use a knife-edge slipstone. This must be extremely fine in order to work in the vee. It is usually available in a 4×1/8×1/32-inch/100×3×0.8-mm size, and often the thicker edge is rounded to serve also as a gouge slipstone.

It is essential that both sides of vee parting tools are honed an equal amount and that the sharpest possible point is formed at the bottom of the vee. Use a flat stone to sharpen the outside.

Veiners are U-shaped and are shar-pened in either two stages, where each side and half the curve is honed, or three, where each straight side is honed, followed by the curve.

Sculpting tools can be brought to an even finer finish by stropping on leather glued to a piece of board. A mixture of flour emery powder, pumice powder or crocus powder and vaseline is rubbed into the leather and acts as a polishing agent. Rub the tool several times along the strop to produce a mirror-like surface on the ground bevel. The tool will move easily over the work and give a finer cut.

The best way to store gouges is in individual pockets in a canvas or baize roll, so the edges don't rub together. To prevent rust and preserve the fine edge, place a small wad of cotton wool in the bottom of each pocket.

Sharpening Vee Parting Tools

1 To sharpen a vee parting tool, the outside is first sharpened on a flat stone. **2** This will create a small ridge at the point of the V.
3 Use a knife-edge slipstone on the inside of the blade to remove this point and the burr.

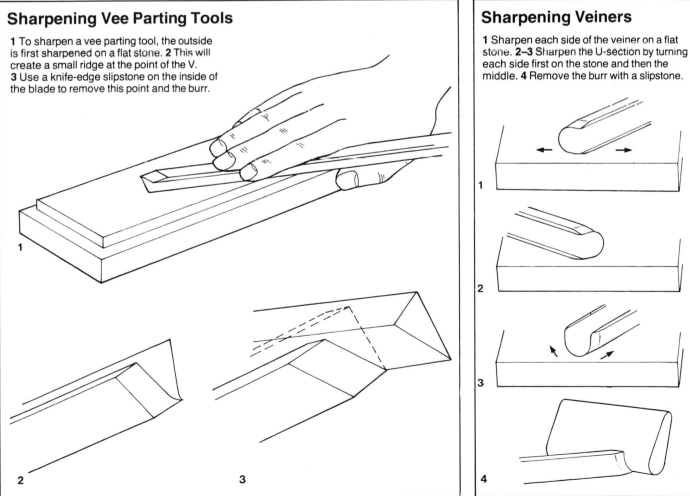

Sharpening Veiners

1 Sharpen each side of the veiner on a flat stone. **2–3** Sharpen the U-section by turning each side first on the stone and then the middle. **4** Remove the burr with a slipstone.

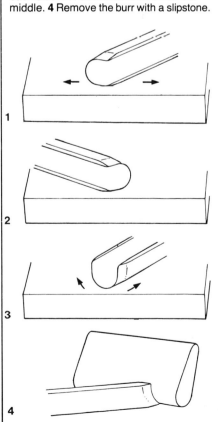

Grinding

Many woodcraftsmen prefer to hand their grinding problems over to a professional, particularly when faced with the cost of grinding equipment. Obviously, it is economical to buy this equipment only if its use justifies the cost.

The earliest grindstones were made of cutting stone, with a fairly soft, open texture. Water was used to cool down the stone.

The use of water wet stone has been reintroduced, on both sides of the Atlantic, as the Sharpenset. Unlike the traditional wet stone, it runs in the horizontal, with a constant jet of water supplied from a small electric pump. It has a fine cutting wheel to which can be fitted a grinding jig for holding jointer and planer blades. This is a completely safe and efficient method of grinding which should have particular appeal to the home craftsman.

The straight dry grinder is powered by electricity, and has one or two wheels, which are either fine or coarse. Take great care when using this tool, as it revolves at a high speed and produces so much heat that it can ruin the tool

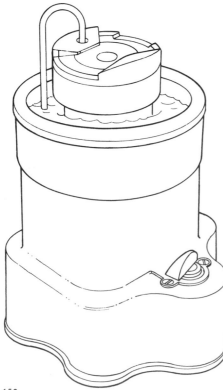

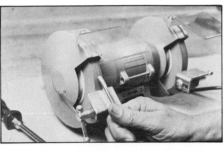

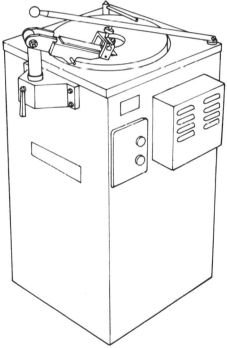

Different types of grinding machines.
Opposite page, far left, a Sharpenset – a water-cooled grinder/sharpener. **Left**, a collection of sharpening tools, including a wet Ohio sandstone wheel. **This page, top**, a 125-mm bench grinder made by Wolf, with different grits at either end of the tool.
Center, a sliding guide made by Elu, which is fitted with a clamp so it can hold a chisel at exactly the right angle when grinding.
Above, a Sharpedge, which is an educational/light industrial machine.

being ground. It must be immersed in water frequently.

New grinders are being produced on the Continent, using man-made cutting grits which are in either a bonded wheel or belt form. These machines can be fitted with rubberized abrasive wheels which absorb some of the heat and make the process safer.

Combination machines have a wheel on one side of a double-ended shaft motor and a drive with an aluminum oxide belt on the other to make grinding easy. Very little heat is generated, and the motor turns away from the operator. One variation has a wet and dry belt working in water.

Perhaps the safest and best of the grinders is the Sharpedge, made by Denford Machine Tools. This machine works in the horizontal, and has an electric motor and oil pump. It is fitted with a tool holder and also a special cone wheel at the side for grinding in-cannel gouges.

A small hand grinder is sufficient for the craftsman who does not do much grinding. Usually, it is fitted with a clamp to fix it to a table or bench. The tool rest is adjustable. Various models are available, with 6×1-inch/150× 25-mm wheels in a variety of grit sizes to suit most needs.

Safety regulations for grinding machines should be read and observed. All machines should be properly guarded and the correct eye protection always worn.

Grinding wheels will eventually become worn, uneven and often badly glazed. Use a dresser to make them serviceable again. There are several types, the most common having a number of serrated wheels of hard cutting steel. The wheels are hooded to deflect sparks and a heavy metal handle is fitted.

Alternatively, an industrial diamond wheel dresser can be used, but these are extremely expensive.

A silicon carbide stick can also dress and re-shape a wheel. This is 6×1×1 inches/150×25×25mm in size and is also suitable for rubberized wheels.

Belts are cleaned with a belt-cleaning stick.

Sharpening Scrapers

Use a burnisher or ticketer to sharpen bench, cabinet and woodturning scrapers. All these tools have a fine steel edge which has to be turned over along its length by a burnisher, to form a burr. Several styles are available, with tapered blades in triangular, round or oval section. They are all fitted with wooden handles.

Bench scrapers are also sharpened on a wheel burnisher. This has a beech body fitted with an accurately-machined guide, and a steel disk mounted centrally in the body burnishes a perfect cutting hook.

When it is new, a cabinet scraper will be square and straight-edged, but after prolonged use it will need to be re-shaped. To do this, place it in a vice and straighten the edge with a smooth file. You can also rub the scraper on an oil-stone to produce the same effect.

The edge will now be straight and fairly sharp and it must be burnished. The burnisher should be steel, harder than the cutter, and fitted with a handle for safety.

Place the cutter flat on the bench and draw the burnisher backwards and forwards across it about twenty times, pressing quite hard but always keeping the burnisher dead flat on the blade. This consolidates the metal.

Now clamp the cutter in a vice and, with the burnisher lying along the top of the cutter, draw it backwards and forwards, pressing quite hard. Gradually lower the handle of the burnisher so that it makes an angle of 15° with the horizontal. This will turn the edge over to form the cutting burr. Repeat this on the opposite side of the cutter.

The cutter of a double-handed scraper is removed from the tool and beveled on both long edges. New cutters are supplied ground at 45° by the factory but they must be sharpened carefully before use. This is best carried out on a fine or medium stone maintaining the same angle. Use a light machine oil on the oilstone and remember to leave the stone clean and dry after use.

When sharpening cutters, always:

1 Keep the burnisher lightly oiled to retain its high polish and to prevent rusting.

2 Draw the scraper blade back into the double-handed scraper before putting it away lightly oiled.

To sharpen the cutter, rest the ground bevel flat on the stone. Keeping the cutter at this angle, move it backwards and forwards across the stone. As the cutter becomes sharp, a slight burr will appear on the flat face and when this extends along the full width, place the cutter flat on the oilstone with the bevel uppermost. A few light strokes up and down the stone will remove the burr.

Always take great care that the back of the cutter lies flat on the oilstone, as any lift will round the back of the cutter and make it useless. It is an advantage if the corners of the blade are very slightly rounded.

To burnish the cutter, place it flat on the bench, bevel side downwards and, using a burnisher of the correct steel, draw it along the edge backwards and forwards about thirty times, pressing quite hard but always checking that the burnisher lies flat on the blade. On no account must the edge be turned over.

Now, set up the cutter in a vice with the ground edge facing you and, with the burnisher lying on the 45° bevel, and pressing quite hard, burnish forwards and backwards. As the burnishing proceeds, gradually raise the handle until the burnisher makes an angle of 15° with the horizontal. This may take up to 36 strokes, depending upon the pressure of the burnisher. A lubricant will make this easier. The edge will now have a definite burr and is ready for use.

To re-sharpen the cutter, remove the burr from the flat side of the cutter either on an oilstone or with a dead smooth file, taking care not to create a bevel. If necessary, rub the cutter on the edge of the oilstone to straighten it.

File or grind the edge to 45° if necessary, and then sharpen as for a new blade.

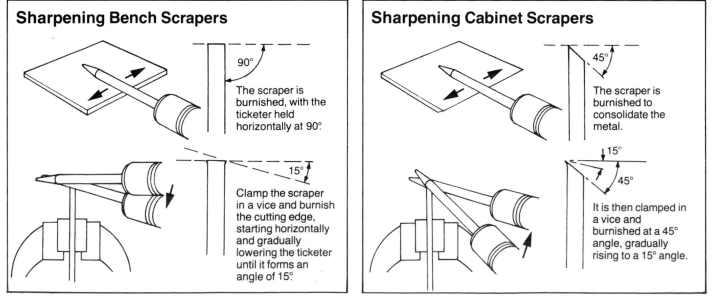

Sharpening Bench Scrapers

90°

The scraper is burnished, with the ticketer held horizontally at 90°.

15°

Clamp the scraper in a vice and burnish the cutting edge, starting horizontally and gradually lowering the ticketer until it forms an angle of 15°.

Sharpening Cabinet Scrapers

45°

The scraper is burnished to consolidate the metal.

15°

45°

It is then clamped in a vice and burnished at a 45° angle, gradually rising to a 15° angle.

Sharpening Boring Tools

The sharpening of auger bits and other boring tools not only requires skill but the correct shape of sharpening tool. Buy a specially-designed auger bit stone, which is shaped to hone the spurs and cutters without making them too sharp and coarse, and abrading other parts of the boring tools.

When sharpening any bit, always observe the following vital rules:

1 Remove as little metal as possible, and do not over-sharpen.

2 Carefully maintain the shape of the bit and never sharpen the sides, or they will bind when boring.

3 Never try to sharpen a bit by grinding it.

Generally speaking, Jennings bits are sharpened more often than necessary and the lives of many are considerably shortened by incorrect filing. Before sharpening, examine a bit carefully to see its profile, and try to keep this when filing.

Both lips of the bit should be filed equally. File them lightly with a smooth file or stone, removing as little metal as possible.

To sharpen the cutters, hold the bit with its screw nose resting on the bench and file the cutting edges on the underside only, ie with the file working through the throat of the bit. It is essential that the cutters are of the same height so that they cut chips of equal thickness.

To sharpen the spurs, hold the bit, nose uppermost, with its twist held firmly against the edge of the bench. File the inside of the spur. You should never file the outside as this reduces the clearance and causes binding and clogging when boring.

The cutter of a Scotch nose bit is sharpened in the same way as that of the Jennings nose bit.

Rest the bit on the bench with the screw lead facing downwards to sharpen the side wings. Work the file through the throat of the bit.

Take great care when sharpening lamp standard augers not to remove too much metal at the nose.

Hold the auger vertically in a vice. Using a dead smooth flat or knife-edge

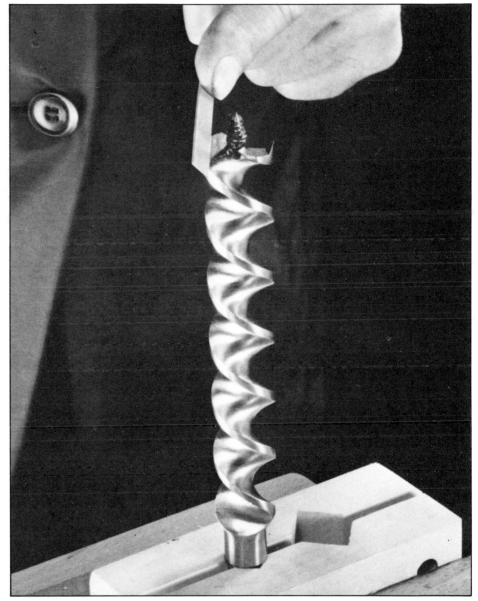

file, make two or three strokes flat on the lip of the nose. Usually, this is enough to restore the knife edge. Check that the center clearance is not obstructed – if necessary, draw the knife-edge file through to remove any burrs which have formed.

To sharpen the shell edges, hold the auger horizontally in a vice. Lightly oil a slipstone of a suitable radius, and move it forward inside the shell, twisting it to the left so that the leading edge of the auger is whetted.

Sharpen a flat bit by placing it in a

When sharpening an auger bit, support it in the vice by placing it in a special block which will hold it securely.

split block held in a vice. Use a knife-edge dead smooth file or a small triangular stone to sharpen the forward cutting edges, maintaining the original angle and keeping the ground surface flat. Both edges should be sharpened equally. Sharpen the brad point if necessary, but maintain the centricity of the point by sharpening both its sides equally.

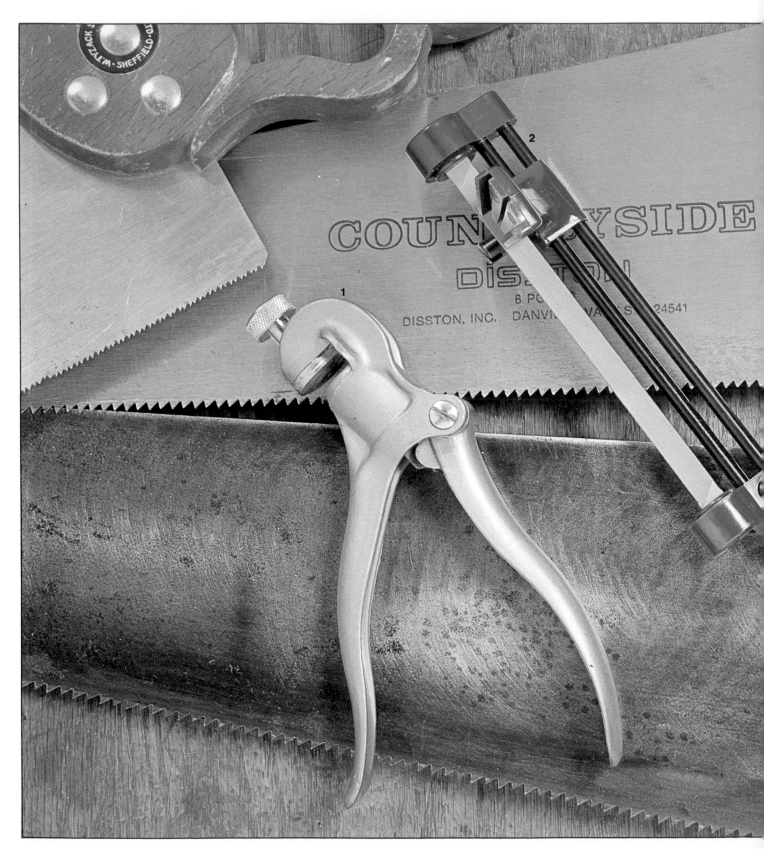

Sharpening Saws

A saw which has been in use for a long time or been abused will need topping, shaping, setting or sharpening.

However, if it is treated carefully, a saw will have been sharpened several times before any or all of the first three processes become necessary.

Once mastered, saw sharpening is a rewarding task – nothing works better than a well-set and sharpened saw. Correctly selected equipment will simplify the job and produce first-class results.

Topping or Jointing: Hand and tenon saw teeth become irregularly-sized after prolonged use. They have to be leveled down to a regular height before they can be sharpened and set.

To do this, place the saw in a vice and fix a file in a grooved clamp. Run the file several times along the complete length of the saw blade. With badly-worn teeth, it may be necessary to shape and top again in order to retain the correct teeth spacing. Remember to keep the file flat by using a topping clamp.

Shaping: Files for sharpening saws are specially made. In the past, they were double-ended. Today, they are triangular, and are supplied without a handle in a variety of sizes. The regular-sized file for the rip saw is 7 inches/175mm long. The 7-inch/175-mm and 6-inch/150-mm extra slim and double extra slim are ideal for saws with 11 to 14 points per inch/25mm. Use the 4-inch/100-mm extra slim for smaller teeth.

Usually, larger saws and log saws are sharpened with a file which has its thicker edge rounded over so that round gullets can be filed. These files have single cut bastard teeth.

To shape the teeth, choose a slim taper saw file with a width of about twice the depth of the saw teeth.

Place the saw in a saw vice, with only the teeth showing. Place the file between the teeth (ie in the gullet), and

Two extremely useful tools for sharpening saws.
1 The traditional saw set, which grips the saw and sets the tip of each individual tooth.
2 A revolutionary tool which accurately sharpens cross-cut, tenon, fleam and rip saws.

Topping

Before beginning to sharpen a saw, it is necessary to support it. The simplest way is to hold it between two strips of wood clamped in a vice, as shown **below**.

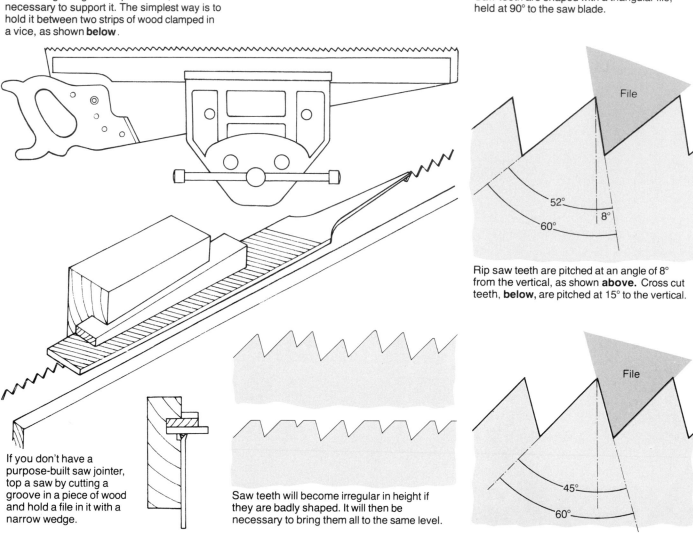

If you don't have a purpose-built saw jointer, top a saw by cutting a groove in a piece of wood and hold a file in it with a narrow wedge.

Saw teeth will become irregular in height if they are badly shaped. It will then be necessary to bring them all to the same level.

Shaping

Saw teeth are shaped with a triangular file, held at 90° to the saw blade.

Rip saw teeth are pitched at an angle of 8° from the vertical, as shown **above**. Cross cut teeth, **below**, are pitched at 15° to the vertical.

press down firmly with the left hand so that the file takes up the correct pitch of the teeth. Draw it straight across the saw at right angles to the teeth. File each gullet carefully, ensuring that each tooth when filed is of the correct shape and height. Any flattening caused by topping will be removed.

Setting: Before any saw can work correctly, its teeth must be set alternately to the left and right. This must be done accurately, so that the points of the saw cut a kerf slightly wider than the thickness of the blade, and the saw moves easily through the wood.

A saw set will perform this task. It is fitted with a hardened steel anvil and a micro-setting screw which controls the distance at which each tooth can be set. Its trigger-type action makes it easy to use, and means it can be used on hand and circular saws.

Buy a plier-type of saw set if you will only use it occasionally. Adjust the pliers to give the correct set, and place them over the tooth. When you squeeze the handles, the plunger will push the tooth over against an anvil. Set alternate teeth in this way, then reverse the saw and set the alternate teeth which you ignored the first time.

Teeth set by hand may have slight irregularities which must be removed by side-dressing. This is done by running a slipstone lightly along the sides of the teeth with the saw held flat on the bench.

Sharpening: Use a file to sharpen cross cut saws. Position the file to work on the front edge of the first tooth set towards you.

The file will also be working on the back edge of the left tooth, which is, of

Setting

The plier-type of saw set should be adjusted to the number of points per inch/25mm of the saw which is to be set. The teeth are set alternately along one side of the saw, which is then reversed so that the other side can be set.

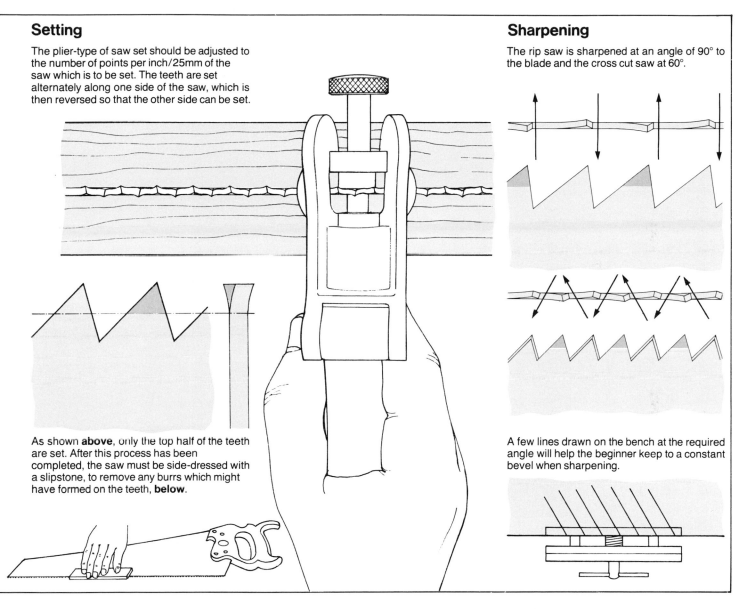

As shown **above**, only the top half of the teeth are set. After this process has been completed, the saw must be side-dressed with a slipstone, to remove any burrs which might have formed on the teeth, **below**.

Sharpening

The rip saw is sharpened at an angle of 90° to the blade and the cross cut saw at 60°.

A few lines drawn on the bench at the required angle will help the beginner keep to a constant bevel when sharpening.

course, leaning away from you. Move the file handle over to the left until it makes an angle of 60° with the saw. Give each alternate tooth several strokes with the file kept at this angle. The front edge of the tooth leaning towards you is then filed each time. Continue filing to the end of the saw, reverse it and repeat the action.

Rip saw teeth are sharpened in the same way as cross cut saw teeth, but the file must be held at 90° across the saw. File each alternate tooth on one side and then file each alternate tooth on the reverse side.

A recent British-designed saw sharpener has two guide bars which keep the file in its correct position in relation to the saw blade. It can be used on saws with between 4½ and 15 points per inch/25mm and is suitable for cross cut, rip or tenon saws.

After sharpening, lightly side-dress the saw with an oilstone to remove any burrs.

Tenon and dovetail saws are sharpened in the same way as cross cut saws, but greater care is needed to maintain the correct tooth size and set. Saws with very small teeth may have to be returned

to the manufacturers for setting and sharpening.

Circular saw sharpeners have to true the periphery of the saw as well as set the teeth. One such tool accepts blades of between 6–12 inches/150–300mm in diameter, and its cone center takes ½–⅞-inch/12–21-mm holes. Lock the circular saw blade before filing and setting.

Several circular saw sharpeners also file the teeth perfectly. The simplest versions clip onto the edge of a bench. They are adjusted slightly to align the file and control the filing.

Fixing Tools

*To fit hinges, brackets and
other metal or plastic fittings, fixing
tools will be needed;
they must be chosen carefully
before buying.*

M

ANY WOODWORKING projects need hinges, brackets and other metal or plastic fittings before they are finished. You will need to buy fixing tools to do these jobs, and like any other woodcraft tools, they must be chosen carefully before buying.

This group of tools can be divided roughly into three categories – those for hammering, those for screwing, and the actual nails, screws or other fixing components.

Hammers

Of these tools, the hammer is the most abused and the one most likely to be selected incorrectly. As with chisels, there is a hammer for every job and there are as many badly-made hammers as there are good ones. The best hammers have carefully-selected wooden shafts, preferably of ash or hickory. These are oil-shrunk at the head to reduce shrinkage. All heads are double-wedged for security – at least one of the wedges being metal.

A whole new range of hammers with steel or glass fiber shafts has been introduced in recent years. These have very long lives but many woodworkers prefer the feel of wood and object to the extra springiness of steel.

The hammer is used for driving in nails and pins to hold material together, to drive in wedges and, in conjunction with a splitproof chisel, adds weight when cutting mortice and other joints. You will need two types of basic hammer – the claw, which removes nails from wood, and the cross-peen, used in fine work.

The Warrington hammer has long been recognized as the best style of hammer for the craftsman. It has a forged head, carefully ground with a flat peen at the opposite end of the poll for starting nails. It is available in six different weights, although these are referred to by number.

No 00	6 ounces/168g
0	8 ounces/224g
1	10 ounces/280g
2	12 ounces/336g
3	14 ounces/392g
4	16 ounces/450g

The pin hammer is light, with the same-shaped head as the Warrington pattern. It only weighs 3½ ounces/98g, with a much longer and thinner handle. Use it for light work – particularly driving in panel pins, glazier's pins and tiny brads.

The adz eye claw hammer is used for heavy work, for extracting nails and breaking apart cases and packages. This tool takes its name from the rectangular-shaped eye, which is used in the adz. It is available with two different types of claw – one has a distinct curve when looked at in elevation, and the other is straighter and used as a lever to open crates. This is often called a ripping-style hammer.

The claw of any hammer must always be correctly forged to grip the nail. In fact, the best way to test a claw is to drive a 4-inch/100-mm round nail completely through a stout plank, and then remove it by drawing it through, with the head last. When it is stood on its head on a bench, a well-balanced claw hammer will rest on the claw with the shaft at a 45° angle, leaning towards you. This hammer is available in four weights.

No 1	13 ounces/364g
2	16 ounces/450g
3	20 ounces/560g
4	24 ounces/672g

Steel-shafted claw hammers are practically indestructible and are sold in similar weights to the wood-shafted type. Some hammers have leather wrapped around the handle, while others have a rubber or plastic sleeve fixed with an adhesive.

When doing craftwork, nails and pins have to be driven without the indentation of the hammer head showing. There are a number of hammers with removable and replaceable faces ideal for this particular work.

The best-known of these is the Thor, which has a diecast aluminum head to which can be screwed two sizes of faces – 1⅜ inch/35mm and 1¾ inch/43mm, made either of rubber or nylon. The head is filled with lead shot to add weight and reduce bounce. It has a metal or wooden shaft.

You can use a pin push, or brad pusher, instead of the hammer when driving panel pins and small headless steel nails. This tool has a small barrel with a magnetized and sprung plunger. The pin is dropped into the barrel and held by the magnet. Push on the plastic or wooden handle and the plunger pushes the pin into the wood.

This tool is particularly useful when pinning hardboard, as the tough outer skin tends to cause the pin to bend when a hammer is used, but the barrel of the pin push supports the pin during insertion.

The pin push or brad pusher holds pins or small nails magnetically in the barrel. Pressure on the handle pushes the pin into the wood.

Opposite, top to bottom, the cross-peen or Warrington hammer has a flat peen for starting nails. The pin hammer is used for driving pins or tacks. Claw hammers have either a curved claw for gripping nails, or a straight or ripping claw for levering open crates. The Thor hammer has interchangeable soft heads to drive nails and pins without indenting the work.

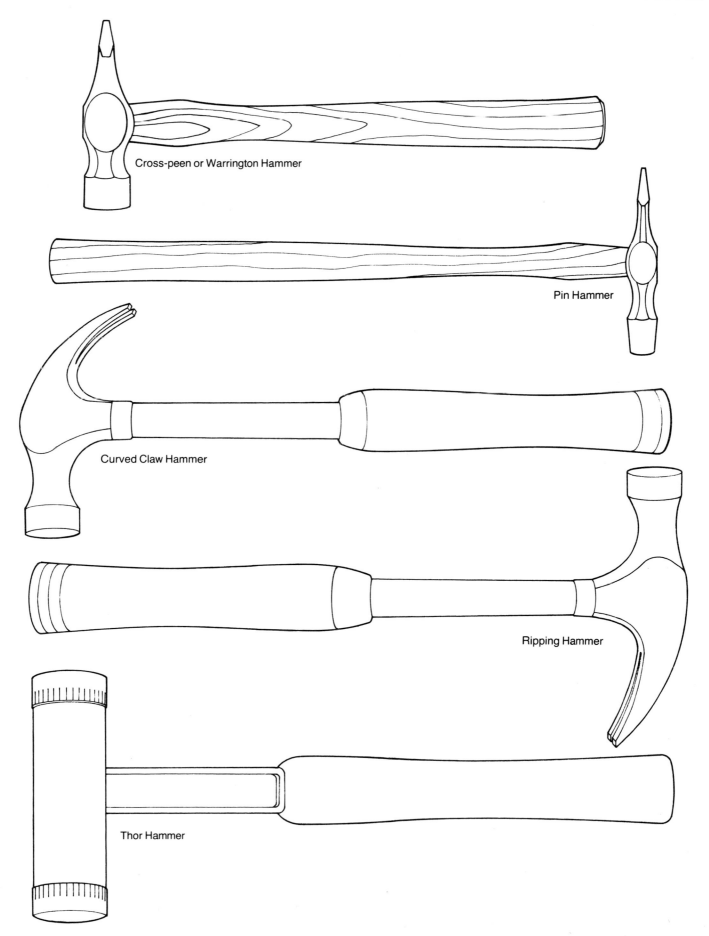

Cross-peen or Warrington Hammer

Pin Hammer

Curved Claw Hammer

Ripping Hammer

Thor Hammer

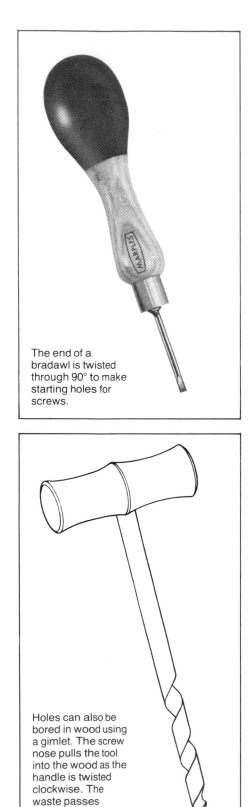

The end of a bradawl is twisted through 90° to make starting holes for screws.

Holes can also be bored in wood using a gimlet. The screw nose pulls the tool into the wood as the handle is twisted clockwise. The waste passes through the shell and out of the hole.

Bradawls

Screws are driven into wood with a screwdriver. Some people start the screw with a hammer – a bad habit which will not only damage the head of the screw, but also split the wood. Usually, the craftsman makes a small starter hole, particularly when hinge-ing. When joining two pieces of wood together, bore the top pieces of wood to clear the shank of the screw and bore a smaller hole in the other piece of wood to receive the threaded part.

Bore these holes with a drill, but form small starter holes with a bradawl. This is a small tool, available with two types of blade and either an ash or plastic handle. The most common one has a round blade, sharpened on both sides to a flat point. Insert the tool into the work with the blade lying across the grain to prevent it splitting. Then twist it through 90° and slowly push it into the timber.

A particularly good bradawl is called the Trigrip, made by Margolis. The handle is styled to fit the hand and give a perfect grip. A lesser-known bradawl is called the Birdcage. The blade is square-sectioned and ends in a point. The square corners cut a tapered hole which is ideally shaped to receive a screw. It is also easier to remove from the hole than the ordinary type.

In the past, bradawls were sized in 1/32-inch/0.8-mm wide steps, but today they are classified simply as small, medium and large.

A tool similar to the bradawl is the garnish awl. This has a beechwood handle and a stout tapered blade. Use it as a point marking tool and to enlarge small holes made with a bradawl.

An older tool which is becoming quite rare is the gimlet. It has a tiny screw nose which runs into a spiral cutter, through which the waste is removed. Originally made of box, the T-shaped handles are now made of beech. Gimlets are available in milli-meter sizes, ranging from 3–7mm. These tools are particularly useful in enclosed situations where a drill, brace or even bradawl would be impossible to use, as they screw into wood easily and without pressure.

Choosing a Screwdriver

A 'madman's paradise' is how a would-be craftsman might describe the range of screwdrivers available. There are more screwdrivers to choose from than tools in any other group. They come in so many sizes and with so many differently-shaped handles that an inexperienced woodworker could be completely baffled by the choice.

However, as in all tool selection, there are certain basic rules to follow. The screwdriver must feel comfortable in the hand, or turning a screw will be a painful process. Always choose a blade that will fit the slot of the screws you are using. Screw heads are no longer just simple slots, but are available in a wide range of shapes, including socket-type heads such as Phillips and Pozidriv, which are cross-shaped.

Screwdrivers can be sub-divided into three styles – standard, ratchet and spiral ratchet.

Standard Screwdrivers

These tools have wooden or plastic handles, with fixed blades.

The cabinet screwdriver is most widely used. It has an oval boxwood, beech or ash handle, styled to fit the hand, and the blade is flattened near the handle so a wrench can be used when additional torque is needed. Cabinet screwdrivers are available in six sizes, with blades ranging from 3–10 inches/ 75–250mm long. All are designed for slot screws.

Another screwdriver designed to fit the hand, but which is specially short for use in awkward places, is the crutch pattern. It usually has a flat oval beech handle with a blade roughly 2 inches/ 50mm long and styled like the cabinet screwdriver.

Many screwdrivers are much abused, particularly if they are hammered. Obviously, they were not designed for this but many stand up to such abuse without damage. However, one screw-driver has been designed to be struck, and is made completely of steel, with side inserts of wood, riveted on to form the handles. It is available 4, 6 and 8 inches/100, 150 and 200mm long, and covers slotted screw sizes from 8–16.

Standard Screwdrivers

Top to bottom, the crutch pattern has an oval handle to fit the palm and is specially short for use in awkward places. The cabinet screwdriver is widely used by woodworkers; it also has an oval handle and a flattened blade when it enters the ferrule to fit a wrench. The London screwdriver has a flat, waisted blade and an oval handle, also flattened on two sides. A screwdriver made completely of steel with side inserts to provide a handle, it is designed to be struck with a hammer.

Right, a parallel screwdriver tip is used to turn screws at the bottom of holes. The flared tip on the left will jam and damage the work, whereas the parallel tip to the right will turn freely. **Far right,** slotted screw and corresponding screwdriver blade sizes.

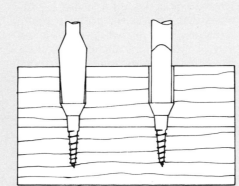

Screw size	Blade size inches/mm	Screw size	Blade size inches/mm
0	$3/32$/2.4	9	$5/16$/7.9
1	$3/32$/2.4	10	$5/16$/7.9
2	$1/8$/3.2	12	$3/8$/9.5
3	$5/32$/4.0	14	$3/8$/9.5
4	$3/16$/4.8	16	$7/16$/11.1
5	$3/16$/4.8	18	$7/16$/11.1
6	$1/4$/6.3	20	$1/2$/12.7
7	$1/4$/6.3	24	$1/2$/12.7
8	$5/16$/7.9		

There is a tremendous variety in the design of handles for screwdrivers. **Left,** square bar, flared tip screwdrivers, with heavy-duty chrome molybdenum blades, plated with zinc to prevent corrosion. Spiral ratchet screwdrivers are sold with a wide variety of drill bits. The Yankee, **below,** has two slotted bits – one Pozidriv bit and one drill bit. **Right,** wooden-handled screwdrivers. **1** These traditional cabinet screwdrivers have polished beech handles. **2** This crutch pattern screwdriver has an extra-wide oval beech handle. **3** A garnish awl used for widening holes already made by a bradawl or for making a hole to locate a drill bit.

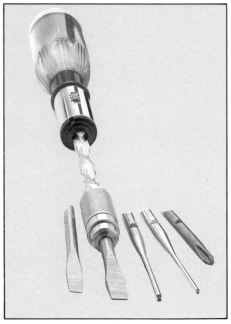

Most plastic-handled screwdrivers have round blades and come in a variety of blade lengths and tip sizes. Choose them with great care. The size of the handle must not only be compatible with the length of the blade, but must also comfortably fit the hand to prevent blistering, while giving maximum turning facility. Avoid molded octagonal or hexagonal handles which give insufficient torque and are extremely uncomfortable to use.

Ratchet Screwdrivers

These invaluable tools were designed to turn screws easily by a simple twist of the hand. They have a simple sliding device which allows the screwdriver to be set in clockwise and anticlockwise directions, and in a fixed position. The better types also have a knurled finger grip which greatly helps in the initial starting of screws. They are available in several sizes and have slotted, Phillips or Pozidriv tips.

Spiral Ratchet Screwdrivers

An extended design of the ratchet screwdriver is the spiral ratchet – sometimes described as the push, or pump, screwdriver. A ratchet device similar to that of the straight ratchet gives clockwise, anticlockwise and fixed movement of the spiral ratchet spindle through a barrel which houses a return spring, allowing the screwdriver to be pushed. It can also be locked in the fully-open position to give a fixed-action screwdriver, providing extended reach and torque. A number of designs have interchangeable bits to cover the variation in types of screw heads. Tiny drill bits are also available. Use this screwdriver when you have a lot of screws to fit, as it will save time and effort.

When working with particularly large, stout screws, a large screwdriver is adequate, but if you own a joiner's brace, buy some turnscrew bits to use with it and make the job easier. The turnscrew bit has one end shaped and flared to suit a standard slotted screw, and the other has a brace tang identical to that on the auger bit, to fit the joiner's brace chuck. It is available in a number of sizes to suit a large range of the bigger-sized screws.

An improvement on the straightforward turnscrew bit is a double-ended pattern, both ends of which fit the chuck and at the same time provide two sizes of bit.

Nails

Nails are most commonly used to fix fittings to wood, and they are available in many forms, all purpose-designed for various crafts. The majority of nails are made from steel wire, with differing carbon contents depending upon the use for which the nails are intended. They are straight or twisted to provide added grip. Some nails are specially hardened to pierce brick or stone.

They are a fast and easy means of jointing timbers together but, as they hold by friction, such jointing is only of limited strength. In many cases, this can be improved by applying glue between the timbers.

The most common nail is the round wire. It has a round shank and a flat head. Usually, the top of the shank is serrated to give added grip, and so is the head to reduce hammer slip. It is available from 1–6 inches/25–150mm long, and is used for rough work – fencing, packaging and other work where it doesn't matter if the nail head shows.

The oval brad between ½–4 inches/12–100mm long, is the nail for the joiner and carpenter. It is oval in shape with a very small head which can be punched below the surface of the timber and hidden with a filler.

A very old nail is the cut nail, also known as the cut brad. This nail is stamped from sheet steel, has square edges and extremely good holding qualities. It is from 1–5 inches/25–125mm long and is used for building work.

A finer round wire nail is the panel pin. This nail is used mainly to secure plywood and hardboard but it can be used to nail thinner timbers. It is from ½–2 inches/12–50mm long, and its diameter is indicated by a number which relates to the Standard Wire Gauge, but the most popular gauges are 18 and 20. It has a very small head.

A similarly-shaped pin is the veneer pin. It is not now in general use, but originally was made to hold veneers until the glue had set.

The escutcheon pin, as its name suggests, is used to secure keyhole plates and other escutcheons. It is made of brass, but is also available as brassed

Nail Types

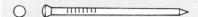

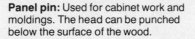

Round wire: The most common nail, which has a round shank and a flat head.

Oval wire: This oval-shaped nail, used in joinery work, must be driven with the head following the wood grain to prevent splitting.

Cut nail: Used to fix floorboards, this nail has square edges and gives a good grip.

Panel pin: Used for cabinet work and moldings. The head can be punched below the surface of the wood.

Veneer pin: It has a similar shape to the panel pin, and is used for small moldings.

Escutcheon pin: A brass or brassed steel pin used to fix such escutcheons as keyhole and door handle plates to wood.

Sprig: A headless tack, tapering to a point, and used to hold glass in window and picture frames.

Tack: Very sharply pointed, it is used to fix carpets and other fabrics to wood.

steel, and has a round shank and a shallow domed head.

A sprig is a headless tack, which is cut from flat steel and tapered to a point. It is the standard nail of the picture framer and is used to secure the back of the picture frame, but can also be used to fix heavy lino to the floor.

Several types of nails are used in upholstery work although now a staple gun is often used instead. It is fast and drives staples into upholstery when a trigger is pressed. It is either hand- or electrically-operated.

Upholstery material is usually held down with cut tacks. These are made in black mild steel, and are very sharply pointed with a flat head. They range from ½–1 inch/12–25mm long.

The brass chair nail has a large head, which is used to cover upholstery tacks, particularly where the sight of a nail head is unavoidable. It is available from ½–1 inch/12–25mm long.

Another nail which has a specific purpose is the gripfast nail for boat work. It looks rather like a screw, but is barbed to counter withdrawal. It is made from silicon bronze, and will neither rust nor stain the timber – a common fault with nails made of iron or steel.

To get the best use from nails, you must always follow certain rules. Choose the correct type of nail for the job, and one of the correct length to give maximum strength. A good guideline is to choose a nail three times as long as the thickness of the board being laid. This is controlled to a certain extent by the thickness of the under timber and whether the nail is to penetrate completely and be turned over.

Stagger the nails – never nail in a straight line as this will split the wood. If there is a risk of splitting, drill a small hole or make one with a bradawl before inserting the nail.

When making boxes, a stronger method of nailing is to slope the nails to give a dovetail effect.

Generally, nail heads are unsightly and when they are to be covered they must be punched below the surface. Choose a nail punch of the correct size to suit the nail head.

Screws

Use screws for a better and more positive method of fixing timber. Screws are stronger, can be easily withdrawn and are superior to nails particularly in temporary constructions. Common screws are manufactured in steel, but brass screws are often chosen for their more decorative appearance and because they don't rust. Use them for timbers like oak, because steel stains the wood as it reacts with the tannic acid.

Screws are made with three styles of head – countersunk, raised countersunk and round. The countersunk head is the one in general use for joining wood to wood, where the head is either flush with the surface or lies slightly below it. The raised countersunk is usually supplied with door furniture and other fittings and can be obtained in a variety of finishes, bronze and chrome being the two most popular. The round head is also used for fittings, usually those of a more general kind such as shelf brackets. They are mostly obtained in black japanned finish, but they can also be brass-, chrome- or zinc-finished.

The screw heads are either slotted or recessed – the latter match a screwdriver with a Phillips or Pozidriv tip. The Pozidriv recess gives better gripping and driving power than the traditional slot.

A screw developed by GKN (Guest Keen & Nettlefold Ltd) in England which is becoming very popular is the Twinfast. Unlike a normal screw, it has two threads. It has a self-centering point which also gives an easy start and keeps its center. It requires only half as many turns and has increased holding power, yet it has a slightly smaller shank which reduces the risk of the wood splitting.

Screws are sold by number and in boxes of 100, 200 and 500, although you can buy smaller quantities. As a rule, the greater the quantity, the smaller the screw. The box label gives the length of the screws followed by a gauge number. This indicates the diameter of the shank of the screw. The shape of head and the material used is also shown.

Prepare the timber properly before inserting screws, to make them easier to work. It is usual with most jobs to bore a hole in the top of the timber which is

Screw Types

Screws are available in brass, steel, copper, aluminum and gun-metal, and have black japanned, galvanized or tin-, nickel-, zinc- and chromium-plated finishes. They are measured from head to tip, and the gauge defined by the diameter of the shank.

Screws have two main types of head – **1** straight and **2** Phillips or Pozidriv – which come in three types of shape – **3** countersunk, **4** raised head and **5** round head.

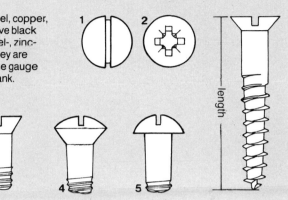

This table gives the size of drill needed to make clearance and pilot holes for each screw gauge. The sizes are given as mm/inches.

Screw	Hardwood		Softwood	
	Clear hole	Pilot hole	Clear hole	Pilot hole
1	2/$^5/_{64}$	1.2/$^3/_{64}$		
2	2.5/$^3/_{32}$	1.6/$^1/_{16}$		
3	3/$^7/_{64}$	1.6/$^1/_{16}$	USE	
4	3.5/$^1/_8$	2/$^5/_{64}$	BRADAWL	
5	3.5/$^1/_8$	2/$^5/_{64}$		
6	4/$^5/_{32}$	2/$^5/_{64}$		
7	4/$^5/_{32}$	2.5/$^3/_{32}$	2.5/$^3/_{32}$	1.6/$^1/_{16}$
8	5/$^3/_{16}$	2.5/$^3/_{32}$	2.5/$^3/_{32}$	1.6/$^1/_{16}$
9	5/$^3/_{16}$	3.5/$^1/_8$	4/$^5/_{32}$	2/$^5/_{64}$
10	5.75/$^7/_{32}$	3.5/$^1/_8$	4/$^5/_{32}$	2/$^5/_{64}$
12	6.5/$^1/_4$	3.5/$^1/_8$	4/$^5/_{32}$	2/$^5/_{64}$
14	6.5/$^1/_4$	4/$^5/_{32}$	5.75/$^7/_{32}$	3/$^7/_{64}$
16	7.25/$^9/_{32}$	5/$^3/_{16}$	5.75/$^7/_{32}$	3/$^7/_{64}$
18	8.25/$^5/_{16}$	5/$^3/_{16}$	5.75/$^7/_{32}$	3/$^7/_{64}$
20	9/$^{11}/_{32}$	5.75/$^7/_{32}$	5.75/$^7/_{32}$	3/$^7/_{64}$

large enough to receive the shank of the screw, and to countersink this hole to receive the head of the screw if a countersunk screw is to be used. A smaller hole is bored in the under-timber approximately half the size of the shank. Don't bore the under-timber if using Twinfast screws.

Drill holes for small screws with a bradawl or gimlet.

Dip the screw in petroleum jelly or grease before insertion to make it easier to work.

Always select the correct length of screw – make sure it is at least three times the thickness of the outer timber. Generally, between six and seven full

threads should engage the wood.

The gauge of screw may be dictated by the fitting being used, but when screwing wood to wood, be guided by the size and thickness of the timber. Small and narrow timbers need small gauge screws.

Decide if you need a particular style of head and finish and whether the screw is to be a decorative feature or not. Remember also to choose the screw which will suit the timber being used.

When using chipboard, run the screw into a pre-bored hole which has been fitted with a fiber or plastic plug.

Placing a screw into end grain is rarely satisfactory, since the threads are held by short grain which can easily be fractured by over-tightening or sudden shock. It is best to insert a dowel crosswise into the timber so that the screw will be held by long grain.

From time to time, you will need special screws and screwed fittings, such as the mirror screw which has a tapped head into which a plated cap can be screwed to cover the head. It is ideal for mirrors and glass panels.

There are various capping devices for hiding screw heads, including one with a snap-on plastic cover, which fits any screw, and another which is tapped into the head of the Pozidriv screw. Both come in several finishes.

A screw particularly useful when fixing a vice to a bench or in heavy constructional work is the coach screw. Pre-bore the wood to be used and choose a suitable tool to fit the head.

Power Tools

*The earliest, and
possibly the most influential, power tool
is the electric hand drill. Today,
few homes are without one.*

Power tools suitable for the home workshop range from the basic electric drill to sophisticated specialist tools. They will perform many tasks quicker and often more accurately than hand tools. Cost, however, is an important factor, and to begin with an electric drill with a selection of the wide range of attachments available may be sufficient. None of the drill attachments, however, do the job as well as specialist power tools.

All electric power tools are potentially dangerous if misused, particularly if the power supply is 230–250 volts. 110 volts, which is the supply in the United States, is safe even when the leads are accidentally severed; in countries where the supply is 230–250 volts, for complete safety use a 110-volt transformer and 110-volt power tools.

Most good-quality branded electric tools are double-insulated – always check that this applies to the power tool you are buying. This feature is so effective that it eliminates the need for an earthing wire.

Buy an extension cable to use with your power tools. They are available everywhere in standard lengths from 25–300 feet/8–100 meters. Buy a heavy-duty cable because it not only loses less electricity, but is also less easily damaged. Keep it on a cable reel, preferably fitted with a plug-in socket, and always unroll the cable completely to prevent a build-up of heat – a strong possibility when the cable is left in the reel.

Another safety measure is to always hold the job in a vice, if it is not already stationary and, if drilling, use the side handle which is supplied with most drills.

The earliest, and possibly the most influential, power tool is the electric hand drill. Today, few homes are without one, although the owners may not be serious woodworkers.

The choice of drills is unlimited, but it must still be made with care. Decide on the size of chuck you need, because they are available ¼, ⁵⁄₁₆, ⅜ and ½ inch/6, 8, 10 and 12mm wide, and with a choice of three speeds – one, two and four. If you are only going to do light work, using

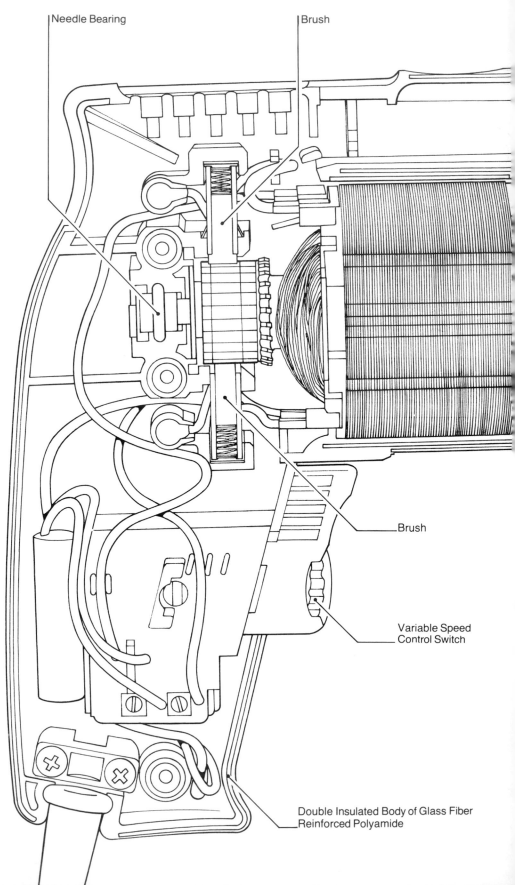

Needle Bearing

Brush

Brush

Variable Speed Control Switch

Double Insulated Body of Glass Fiber Reinforced Polyamide

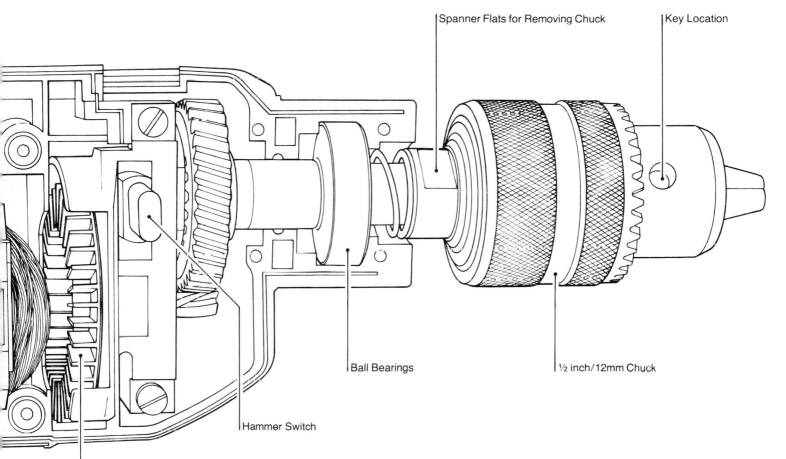

Spanner Flats for Removing Chuck

Key Location

Ball Bearings

½ inch/12mm Chuck

Hammer Switch

Fan

This drill has a speed control variable between 0 and 3400 rpm. It also offers a choice of nine constant speeds, and is especially useful for tackling brittle surfaces such as tiles or glass.

small wood drills and flat bits, a ¼-inch/6-mm wide chuck with a single-speed drill is adequate, although it is useful to have a choice of two speeds, such as 1000 rpm and 2500–3000 rpm if drilling steel or harder woods at low speeds.

If you use machine bits, you must have a drill with a ½-inch/12-mm wide chuck and a choice of speeds because the bits have a ½-inch/12-mm wide shank and must bore large holes.

All hand electric tools comprise a basic electric motor with an extended spindle, which connects with a drill chuck and various attachment devices in other tools. The chuck usually has three jaws and is operated by a simple chuck key. The tool is held in the hand, usually by a pistol grip, and the motor is started by pressing a trigger. The power supply is cut and rotation stopped when the trigger is released, but by pressing down on a small button the trigger can be kept in the 'on' position without pressure. A

fan device built into the machine keeps the motor cool.

A choice of speeds is effected either by a two-speed gear reduction or by an electrical device known as a diode switch. Unfortunately, if you use the latter, you will have a definite loss of torque at lower speeds. Another method, which is becoming more popular, is built into the drill trigger, where the speed is altered by varying the finger pressure.

If you want a heavier drill with a back handle, make sure that the speed will suit your needs, since many of these are slow-speed machines.

Most craftsmen with a small workshop will need a vertical drill press. There are several different types available, from the simplest stands to versions fitted with depth stops. If you are planning on doing a lot of morticing, you should buy a drill with a morticing attachment. These come with a specially-devised attachment to the

A mortice stand is shown **right**, with the mortice attachment in close-up **above**.

stand which holds the shank of the square chisel while the bit runs inside and is motivated by being held in the drill chuck. A clamping device is attached to the base for holding and accurate positioning of the timber. You could also add a depth-setting device with a micrometer-type adjustment. Check that your drill will match the stand – many stands have interchangeable fittings which will take a variety of electric hand drills. Bolt the drill stand firmly to the bench and use a guard.

Few drill stands have drill press vices fitted, but small work must be held safely with the fingers as far away from the moving boring tool as possible. Drill press vices can be bolted to the base of the stand and have capacities of more than 4 inches/100mm. If you do use one, you must protect the wood by placing a small piece of timber between the wood and the base of the vice.

If you are using a drill stand with small work, it is important to use a drill press vice.

Hand Power Saws

The circular sawing of wood uses a great deal of power, so it is essential to keep the saw in perfect condition to obtain the best results. The cheap hand power saw and saw attachment can both be poor investments. However, a well-chosen saw can be quick and accurate. With correctly-selected saw blades, it can cut a wide variety of materials, and when fitted with abrasive disks, its range is extended to cut such articles as brick, concrete, slate and asbestos.

There are three variations of saw, the most common of which is the hand-held, which has a blade with a diameter from 5–9 inches/125–225mm. The blade is vertically mounted on the shaft of the motor and secured with a nut, and the attachment is fitted with a sole, on which the saw runs, a fence and an automatic saw guard. This moves around the saw when sawing but automatically drops into place for safety whenever the saw is removed from the timber. The sole can be tilted for angular cuts. Fitted with the normal trigger, it also has a button which locks it in continuous rotation.

A number of manufacturers make a saw table, to which a power saw can be fitted. Usually, the saw is attached by screws through its sole to the underside of the table. In the upward position, the blade passes through the table. These tables are usually fitted with a fence and riving knife exactly as the commercial circular saw. Both hands are free to push the wood through the saw, but always use a push stick to keep your fingers well away from the blade.

The third variation converts the saw to a radial arm saw. This is an attachment for the saw which comprises a sliding arm on the pillar of the drill stand. Place the saw on a purpose-built wooden table. It slides along its horizontal arm, and can be set at any position along the wood. This is certainly not as sophisticated as the popular radial arm saw, but is adequate for limited use.

A general rule for circular saws such as the one shown **left** is that smoothness of cut increases with power and speed. You should check how far speed is maintained under load. Slow down of about two-fifths under full load is normal. Kickback is eliminated in modern tools by use of a slip crutch. **Above,** the jig saw with its short narrow blade is designed for enclosed cutting, cutting into wide sheets and cutting curves.

Right, a circle-cutting guide can be attached to the rip fence to cut perfect circles. **Below,** the sole plate of the jig saw can be adjusted to cut a bevel. It is important to avoid forcing the saw, to keep its base flush with the surface of the work, and to watch out for the edge of the blade touching clamps holding the work.

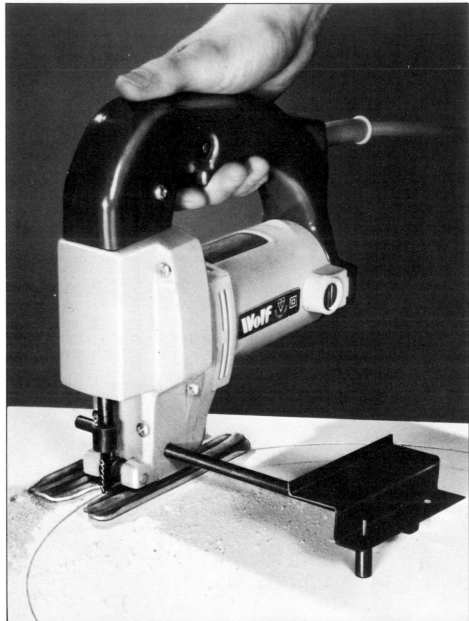

Choose the saw attachment with care and make sure that not only will it fit your drill, but that it will not need more power than your drill can provide. The size of the center hole of the saw blade must match that of the drill chuck arbor, and the body of the drill must, of course, accept and hold the body of the attachment securely.

Another extremely useful saw, ideal for cutting curves and also cutting into wide sheets, is the jig saw. This is avail-

able with an integral motor or as an attachment to a power drill. Once again, it consists of a simple electric motor, with a spindle connected to a cam mechanism to give an up-and-down reciprocating motion to a short, narrow blade. Some models have two speeds or one variable speed. In some machines, a tube directs air from the motor to behind the saw blade to blow fine saw dust away from the work. There is a limit to the thickness of the wood to be

cut, but most machines can cut wood up to 2 inches/50mm thick. Slowly move the tool into the wood, and don't cut curves beyond the capacity of the blade – it will not turn right-angled corners. For enclosed cutting, such as circles within a sheet, tilt the jigsaw forward onto its nose, and gradually lower the blade until it passes through the timber, with the machine finally resting in the vertical position. If you decide to buy a drill attachment, make sure that it fits.

Choosing Saw Blades

There are various circular saw blades available, all derived from the hand saw but further extended to provide different types of cutting.

The rip saw blade has coarse teeth designed to cut timber lengthwise with the grain.

The cross cut blade has smaller teeth for cutting across the grain and for panel cutting.

The combination tooth blade will cut in any direction and at any angle.

A well-thought-out design of combination teeth provides a planer blade which gives a very fine smooth edge.

Most blades are available in Teflon or other coated finishes. These mean the saws move through the timber easier, keep their edges longer and do not rust or corrode.

Buy a tungsten carbide-tipped blade if you intend to cut a lot of man-made boards, such as particle boards. These saws have been developed because the grit content, combined with the hard glue and the variation in hardness of the timber chips, quickly blunts the teeth of a normal saw. The blades are expensive and, in spite of their hardness, the teeth are rather fragile and have to be handled carefully.

Circular Saw Blades

A wide variety of circular saw blades is available, making it possible to cut most materials. Blades can be Teflon-coated for prolonged sharpness and to prevent rust, with a black oxide finish, or tipped with tungsten carbide, which is especially hard but should not be used on stone, masonry, ferrous metals or wood with nails.

The combination blade is capable of cross-cutting, rip-cutting and mitre cutting.

The cross-cut and flooring blade makes a smooth cut across the grain.

Plywood and paneling blades have very fine teeth for a clean finish on plastic, chipboard and soft metals as well as plywood.

The hollow ground planer is useful on cabinet work and will give an especially smooth finish on all solid woods.

Metal slicing blades will cut light gauge metals, transite and fiberglass.

Jig/Saber Blades

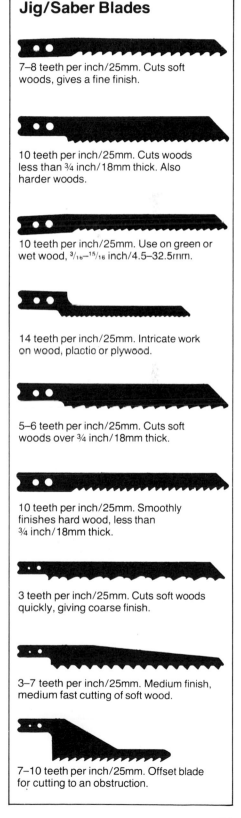

7–8 teeth per inch/25mm. Cuts soft woods, gives a fine finish.

10 teeth per inch/25mm. Cuts woods less than ¾ inch/18mm thick. Also harder woods.

10 teeth per inch/25mm. Use on green or wet wood, ³/₁₆–¹⁵/₁₆ inch/4.5–32.5mm.

14 teeth per inch/25mm. Intricate work on wood, plastic or plywood.

5–6 teeth per inch/25mm. Cuts soft woods over ¾ inch/18mm thick.

10 teeth per inch/25mm. Smoothly finishes hard wood, less than ¾ inch/18mm thick.

3 teeth per inch/25mm. Cuts soft woods quickly, giving coarse finish.

3–7 teeth per inch/25mm. Medium finish, medium fast cutting of soft wood.

7–10 teeth per inch/25mm. Offset blade for cutting to an obstruction.

Power Plane

Another tool which is available with integral motor or as an attachment for a drill is the power plane. It can be a very useful addition to a power tool kit and is used not only for straight planing but also for rebating and chamfering. With adaptation, it can be used to cut tenons. The planer consists of a motor, which in turn rotates a small drum. Two or three small blades are accurately fitted to the drum and slightly raised from the surface of the drum. The plane can be adjusted to varying thicknesses of cut and is fitted with two handles like a normal bench plane. The blades can be sharpened in the machine. Always disconnect the machine when making adjustments and secure the timber in a vice before beginning work.

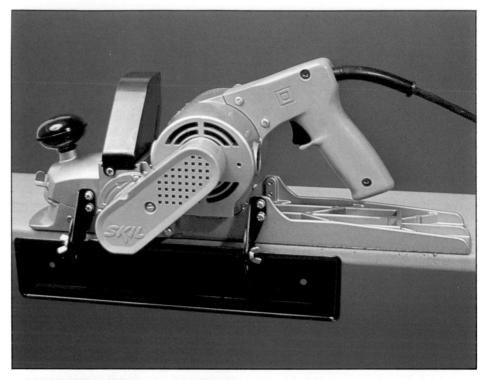

The power plane can be used for rebates and chamfering as well as normal planing. The version, **right,** has an adjustable depth of out of 0–3.2mm and a reversible dust and wood shavings deflector.

The orbital sander, **facing page,** has a suction attachment for removing dust. The palm sander, **right,** is useful for lighter work. The belt sander, **below,** is a valuable acquisition if you envisage a good deal of heavier work.

Sanders

You will probably use flint or abrasive paper for many jobs, and unless you are sanding large areas, abrasive paper wrapped around a cork block will suffice. However, apart from woodcraft work, there are many other jobs around the home which need rapid sanding. For an odd job, you can fit a drill with a rubber backing disk which has been faced with an abrasive paper. You must take great care not to score the wood, or even burn it, if you use this method, and if you intend to do a lot of this work, buy an orbital sander.

This machine was first introduced 25 years ago and was greeted with instant success. It consists of a flat, rectangular sanding pad to which is clamped a sheet of abrasive paper. The clamping clip keeps the paper in tension and in contact with the felt or rubber backing sheet attached to the pad. The motor transmits a very small orbital movement which gives the required finish, depending on the size of grit on the abrasive paper used. These sanders appear as attachments for the electric drill or as self-powered units.

If you don't intend to use the sander very often, buy one as an attachment, but if you will use it a lot, buy a separate machine which has a vacuum dust collection bag fitted.

All these machines are fired with a trigger and lock-on button. Always secure the work first so that you can use both hands on the machine.

An alternative is to use the drum sander. These are revolving drums of plastic foam or rubber, with an arbor which fits into the drill chuck. An abrasive belt is fitted around the drum. This machine is suitable for small work, cleaning up curves and edgings, but shouldn't be used for large surfaces.

A more powerful tool is the belt sander, which sands both wood and metal very quickly. It also finishes timber or metal and will remove old varnish or paint.

The abrasive belt is fitted over two rollers to give a continuous sanding action. A pad lies between the rollers and holds the belt flat against the work being sanded.

Routers

Already widely used in North America, routers are now gaining in popularity all over the world.

There are several styles available, but basically the machine consists of a motor with either a simple chuck with the round cutter shank held firmly with a grub screw, or a collet chuck, fitted to the shaft. The tool has depth setting adjustment and a fence for straight or curved cutting. Most routers can also be fitted with guides for use with templates, as well as a trammel attachment for circular work. Dovetail cutters used with a dovetailing attachment simplify the jointing of board in cabinet work.

In a normal router, the cutter is set to the required depth, and the machine is then switched on and lowered or pushed into the timber. A better router is the plunging type which allows the cutter to be gently lowered into the work and fixed at an agreed depth. When the cutting is finished, it can be withdrawn instantly into the machine housing with the machine still standing on its base.

The router rotates at speeds of up to 30,000 rpm – much faster than any other power tool. The wide range of cutters are precision-made in either high speed steel or tungsten carbide and, in combination with high revolutions, cut perfectly smooth grooves, rebates, moldings and curves in variety. The machine can also be used for cutting housings (dados), dovetails, mortices, tenons and tongues. The woodcarver can also use the tool to quickly remove ground wood or to carve. If you fit it with the correct cutter, you can use it for lettering.

Some manufacturers make a special table to which the router is attached. This effectively converts the router to a spindle molder, because the cutter passes through and projects above the table. Fitted with a fence, it can be used most effectively when making moldings, rebates and doing ploughing and joint cutting. You can easily make one to suit your own router.

With the right attachment, a router like the one **right** will cut smooth grooves, rebates, moldings, curves and the shapes necessary for various joints.

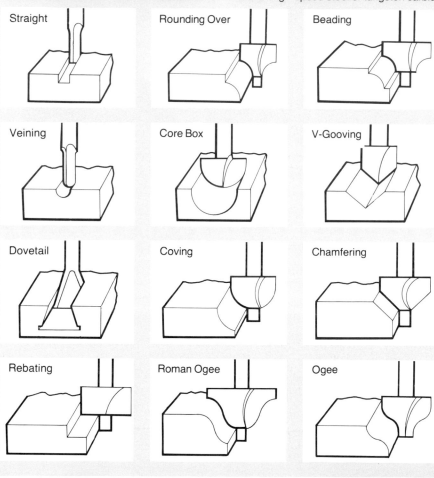

Router Bits

Modern routing machines operate at very high speeds, so the attachments must be perfectly balanced and made of the best materials.

Below are attachments for just some of the jobs for which a router can be used. They are made of high-speed steel or tungsten carbide.

Straight

Rounding Over

Beading

Veining

Core Box

V-Gooving

Dovetail

Coving

Chamfering

Rebating

Roman Ogee

Ogee

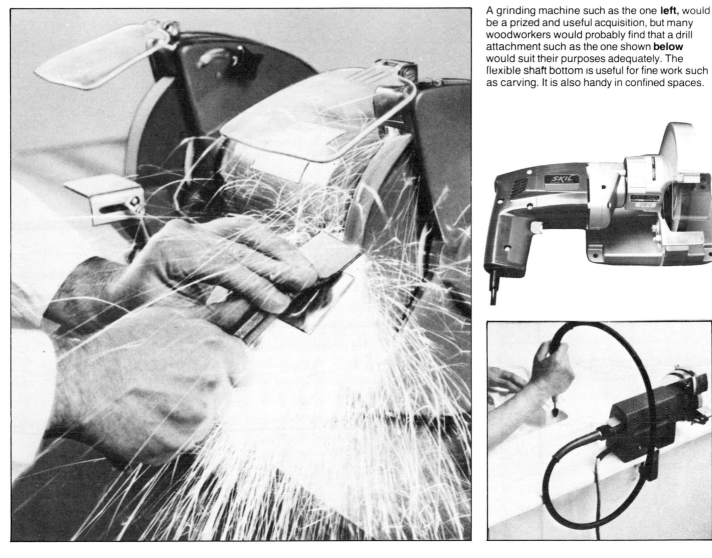

A grinding machine such as the one **left**, would be a prized and useful acquisition, but many woodworkers would probably find that a drill attachment such as the one shown **below** would suit their purposes adequately. The flexible shaft bottom is useful for fine work such as carving. It is also handy in confined spaces.

Bench Grinders

An additional power tool which you might find useful is the bench grinder. Since most people hate sharpening tools and have little or no experience in grinding them, this machine removes a great deal of frustration.

An electric motor is fitted with two shafts, one at each end. Grinding wheels with various grades of cutting grit are attached to the shafts, and the wheels are held in place by nuts which screw down onto a collar on the threaded shaft. The left-hand wheel is held by a left-handed screw and nut to prevent it working loose while in use.

Similarly, the wheel on the right is right-handed. Each wheel has a spark guard and an adjustable tool rest to take up wheel wear, and both are shrouded to prevent accidents. These tools are supplied with one fine and one coarse wheel, usually of manufactured grit. One or two European grinders sell rubberized abrasive wheels, and these are superb for grinding and sharpening, as they have a very gentle action even when the coarse wheel is used, and the bevels polish the wood to make it easier to cut. These wheels will revolutionize tool sharpening, particularly for the home craftsman and amateur.

As an alternative to a bench grinder, use a horizontal drill stand in conjunction with a power drill. Most of these stands are available with a grinding wheel attachment. The stand must match the drill and be screwed down firmly to either a bench or a board.

Flexible Shafts

The flexible shaft is a useful accessory for the power drill. It is merely an extension of the drive through a flexible shaft to another chuck into which can be fitted flap wheels, small cutters, drills and a host of other cutting aids. This tool is particularly useful when working in enclosed spaces and so is suitable also for the carver and sculptor. The more expensive shafts can be fitted with various attachments and tools, and they can also run from an electric motor, drilling machine, grinder or lathe.

Glossary

Adz: An axlike tool, with the blade set at right angles to the handle and curving towards it. It was used by many wood-craftsmen before the plane came into general use. The timber was set up on the floor. The hand adz is still a very popular tool amongst carvers and sculptors.

Aloxite: Trade name for a man-made abrasive used in grinding wheels and sharpening stones.

Aluminum Oxide: A long-life grit for abrasive wheels and papers.

Allongee: Large taper blade chisels for carving.

Angle Divider: A tool used to transfer angles from one piece of timber to another, with an infinite variety of settings.

Ash: Similar to hickory in color and qualities. Bends well when steamed.

Awl: A type of marking tool.

Back Iron: *See* Cap Iron.

Back or Tenon Saw: Used for small work on the bench top.

Banding: Thin, narrow strips of different colored timbers made up in patterns and inlayed into grooves cut in cabinet work as a decorative feature.

Batten: Square strips of timber of varying size.

Bead: A semi-circular cut in a molding.

Beading Saw: *See* Gents Back Saw.

Beech: Light, reddish-brown timber, with straight, fine, close grain. Fairly hard. Very popular for tools and furniture generally.

Bench Hook: A board which hooks over the bench top with an upstand against which timber is placed for sawing.

Bench Stop: A wood insert, passed through the bench top close to the left-hand end, against which timber can be held when planing.

Bevel: A tool used for setting off angles, as in dovetailing.

Blitz Saw: A small saw with a thin blade designed for cutting metal but which can be used for wood and plastics. It has a small hook at the end for the left hand.

Block Plane: A small plane used on end grain.

Bow Saw: A narrow saw, set in a wooden frame, for cutting curves.

Boxwood: A hard close-grained timber, ideal for chisel handles.

Brace Tang: The part of the boring auger which is shaped to fit the chuck of the joiner's brace.

Brogue: Scottish name for a bradawl.

Bull Nose Plane: A plane for cutting close into a corner. The cutter is very near the front of the plane.

Burnisher: A hard steel rod used to turn over the burr on scraping tools.

Burl or Burr: An excrecence found on many trees, usually formed around an injury to the trunk. Valued for veneers.

Butt Chisel: A short-bladed chisel used when cutting housings for butt hinges.

Butt Gauge: Used to mark out for hinging.

Butt Joint: Two pieces of wood joined lengthwise without overlap or tongue.

Calipers: Both outside and inside – used for measuring curved work.

Carcase: A skeleton or frame of a piece of cabinet work.

Carver's Hook: For scooping out curved recesses.

Carver's Mallet: A round headed mallet specially designed for the carver.

Carver's Screw: A screw turned into the base of a carving, passed through the bench and screwed down with a wing nut to hold the carving securely.

Cast: Timber twisted in its length is described as cast.

Cap or Back Iron: The curved iron which attaches to the plane iron to control the cut.

Caul: A metal plate used in veneering. It is warmed and applied to the veneer to keep it in position while the glue is setting.

Cellulose Acetate Butyrate: A hard splitproof plastic used in quality chisel handles.

Center Square: A tool used to locate the centers of circles.

Center Punch: Used to mark centers for drilling or boring.

Chamfer: Cutting a square edge equally on both sides to form a bevel.

Chatter: Caused by a badly set plane cutter.

Chip Carving: Shallow carving in straight lines and geometric forms.

Choke: Shavings which fail to pass through the mouth of the plane and out of the escapement will cause the plane to choke.

Compass Plane: Sometimes called circular plane, it has a flexible sole which can be set for convex or concave cutting.

Clamp or Cramp: A tool which holds pieces of wood together.

Collet Chuck: A chuck with a simple collet to grip small round stock on the lathe.

Collar Chuck: Used for mounting timber on the lathe for egg cups, etc.

Combination Plane: A plane capable of ploughing, rebating, beading and tonguing.

Compass Saw: A thin taper bladed saw for cutting tiny holes.

Concave: Hollowed out.

Cone Center: Mounted on the headstock for driving square material without centering.

Convex: Bulging out.

Copying Attachment: A device to set the lathe for repetition work.

Corrugated Base: A series of longitudinal grooves cut in the soles of bench planes to relieve the suction between the wood and the sole of the plane, giving easier cutting.

Cove: A concave cut.

Crotch: A veneer which has been cut from a limb crotch.

Cross Cut: A saw for cutting across the grain.

Curl: Feather grain found in timber, mainly mahogany.

Dado: A groove or housing made across the grain.

Deal: A word used loosely to describe timbers cut from pine or fir logs.

Dog: A peg inserted into a hole in the bench top and used in conjunction with a tail vice to hold timber securely.

Dowel: A round pin or peg used in jointing.

Draw Knife: A two-handed tool much favored by the cartwright and wheelwright for shaping and chamfering. There are many variations available.

Drill Chuck: Used in the wood lathe tailstock to hold machine boring bits.

Edging Tool: For trimming veneer.

End Grain: Timber grain which shows when a piece is cut transversely.

Expanding Collar Chuck: A six-in-one chuck which eliminates the need for faceplates, collar chucks, screw chucks, etc.

Face Side and Face Edge: The accurately planed face and edge of a board or batten from which all marking out is done.

Fence: A part of a tool designed to limit movement, as in the fence of the plough plane.

Fiddleback: Design of a chairback. It also refers to the grain of timber which shows a rippling effect. Very much prized in the backing timber for violins.

Figure: The formation of grain.

Fillet: A strip of timber added to work either as a guide or a support.

Fillister: A rebate plane used in cutting sash windows and other work.

Finger Plane: A small plane used by the violin-maker.

Flapwheel: Flap of abrasive paper, attached to a center core and used in an electric drill.

Flat or Spade Bit: Designed for use in electric drills.

Fleam: The teeth of a cross cut saw, which are shaped like an isoceles triangle. Ideal for quick cutting.

Flooring Saw: A special saw for cutting flooring – useful to the electrician in particular.

Fluteroni: Wide U-shaped carving gouge with round corners for shaping round sides.

Fore Plane: A plane between the trying plane and the jack plane in length.

Forstner Bit: An auger bit which runs on its periphery, used exclusively for boring shallow, flat-bottomed holes.

Frame Saw: A saw with a narrow blade tensioned and supported in a wooden frame.

Frog: The device in a plane which holds the cutter in place and which can be adjusted to control the cutting.

Gain: A recess to receive a butt hinge.

G Cramp or Clamp: So-called because it is similar in shape to the letter G. It is used for small cramping work.

Geelim: The Scottish name for a shoulder plane.

Gents Back Saw: A small saw with round handle and thin blade, used for dovetail cutting and small work.

Green Wood: Unseasoned or newly felled timber.

Grinding: Resharpening badly-worn or -shaped cutters by holding them against a rotating wheel.

Grounding: Cutting away the background in a low-relief carving.

Hardwood: Generally the timber of deciduous trees.

Heartwood: The best timber, which comes from the heart of the tree and has matured with age.

Heat Treatment: The hardening and tempering of cutting steels to ensure long-lasting cutting edges.

Hickory: Whitish-yellow in color, very tough hard elastic and strong, ideal for hammer handles.

Hogging: Rough planing of timber, usually with a jack plane that has a slightly rounded cutter.

Holdfast: A clamping device, dropped through a hole in the bench top and used to secure timber for carving and other work.

Honing: The fine sharpening of a cutting edge using a fine sharpening stone, called a hone.

Honing Guide: A device to hold chisels and plane cutters at a constant angle during sharpening on an oilstone.

Horn: The forward handle fitted on the older planes.

Housing or Dado: A groove cut across a board to receive a shelf.

In-Cannel: A gouge ground on the inside.

Inlay: A form of decoration where pieces of wood and other materials are set into a base of wood and left flush.

Inshave: A tool similar to the draw knife but which has a curved blade for hollowing out.

Iron: The cutter of a plane or spokeshave. These were originally made of iron with the cutting steel face welded on.

Jack Plane: Takes its name from 'Jack of all Trades'. It is the most commonly used plane.

Jointer: A long plane, particularly useful when planing long boards for rub jointing.

Keying: An addition to a piece of wood used to strengthen it. Dovetail keying is where the piece of wood is inserted into a dovetailed housing to prevent a board from warping.

Keyhole Saw: Similar to the compass saw, having a thin tapering blade for cutting keyholes and similar slots.

Knot: The very hard wood at the junction of a branch, sometimes completely circular and loose.

Lengthening Bar: An addition to a sash cramp to give added capacity.

Lignum Vitae: The tree of life – it produces an extremely hard wood with an oil secretion useful for bearings. Highly prized for small planes and plane soles.

Live Center: *See* Running Center.

Log Saw: A thin-bladed saw with a tubular frame used for logging and large rough work.

Long and Strong: Woodturning tools with thick strong blades and matching handles.

Macaroni: Wide U-shaped carving gouge with square corners, designed for finishing the sides of shallow recesses.

Mahogany: Reddish-brown timber, fairly hard, with many varieties. Once used as a base for veneered furniture.

Makrolon: A flexible plastic material used to make modern rules.

Mandrel: A lathe attachment used to mount timber on the lathe for turning.

Mitre: The joining of timber by cutting it at an evenly divided angle – usually 45°.

Mitre Square: A tool incorrectly named, which is used to test and mark lines at 45°.

Morse Taper: Both head and tailstock of the lathe are bored on a universally agreed taper to receive standard components.

Multi-Plane: A plane capable of many types of cuts.

Multispur Bit: *See* Saw Tooth Machine Bit.

Nosing: The front of a stair tread.

Ogee: A type of molding.

Oilstone Slip: A shaped oilstone for sharpening gouges.

Old Woman's Tooth: An early name for the router plane.

Out-Cannel: A gouge ground on the outside.

Ovolo: A type of molding.

Palm Plane: A plane which comfortably fits into the hand. It is useful for the musical-instrument-maker and others handling small work.

Parting Tool: Tool for parting off work on the wood lathe.

Plinth: A receding part of a cabinet next to the floor.

Plough: A plane for making grooves or rebates.

Plug Cutter: A special machine cutter for cutting wooden plugs.

Pocket Chisel: *See* Butt Chisel.

Polypropylene: A plastic material used in the making of tool handles, which is practically indestructible.

Purfling Tool: Violin-maker's groover.

Quick Gouge: A deep carving gouge used for roughing out.

Rail: The horizontal member of a frame or carcase.

Rebate or Rabbet: A cut made on the edge of a frame or board to receive glass or a wooden panel.

Reform Plane: A continental plane with an adjustable mouth.

Registered Pattern: A thicker-bladed chisel with steel hoop and ferrule, often called a mortice chisel.

Revolving Cone Center: To hold square material in the tailstock for turning.

Rifflers: Double-ended rasps or files for the woodcarver.

Rip Saw: A saw with special teeth for cutting with the grain.

Rosewood: Beautiful, dark purple/brown-colored timber used for expensive cabinet work, tool handles and other components.

Router: A hand or machine tool used to make a variety of cuts.

Running or Revolving Center: A center running in a ball race and mounted in the woodturning lathe tailstock.

Sapwood: Timber immediately underneath the bark, still growing and too young for use by the woodcraftsman.

Saw Set: A pincer-like tool for setting saw teeth.

Saw Tooth or Multispur Bit: A machine bit with saw-like teeth around its periphery which is designed to cut shallow or deep holes in any timber.

Scant: The bare measure.

Scopas Chops: A carver's/sculptor's vice, made of wood with a central screw for mounting to the bench top. It has a deep capacity, with leather-protected jaws.

Scorp: A circular-bladed carver's tool.

Scraper: Used in the hand to finish after planing.

Scratch Stock: A small tool used to cut shallow recesses for inlaying.

Screw Chuck: A chuck with single screw for mounting small work on the lathe.

Screwdriver Bit: A brace bit with a screwdriver end.

Scrub Plane: A roughing plane.

Set: The inclination of alternate teeth of the saw to the right and left. Also, the distance the cap iron is set back from the cutter edge in the plane.

Shake: A seasoning split seen in timber.

Sharpenset: A fine wet grindstone.

Shoulder Plane: A precision plane used to trim end grain before jointing.

Silicon Carbide Grit: Used on wet-and-dry abrasive papers. The paper and glue are waterproof, permitting the paper to be used wet.

Sizing Tool: For accurate sizing and setting out on the lathe.

Skew Chisel: For planing wood between lathe centers, squaring, beading, curving and tapering.

Slow Gouge: A shallow carving gouge for finishing work.

Spalted Timber: Timber which is brittle through decay.

Spindle Gouge: Round nose gouge for coving in spindle work on the lathe.

Spokeshave: A cutting tool with two handles used for small curved work. It has a similar action to the plane.

Spurs: The small cutters of the combination, fillister and multi-plane, brought into use when cutting across the grain to sever the top fibres ahead of the cutter.

Stringing: Tiny strips of thin square or rectangular section timber used in inlaying.

Surform: Registered name for a new patterned file.

Tail Vice: A vice at the end of a bench which uses dogs to clamp boards for planing, etc.

Tenon: The male part of the joint which fits into the mortice.

Toat: Another name for the plane handle.

Toothing: The roughing of the surface of timber to provide keying for veneer.

Trammel Point: Used in conjunction with a beam to mark out large circular work.

Trying Plane: A long bench plane used as the jointer.

Tungsten Carbide Grit: The ultimate in man-made abrasive grits.

Vee Parting Tool: V-shaped carving tool for marking out texturing and lettering.

Veneer: Thin timber glued to a solid timber background for decoration, enabling the use of rare and exotic woods not available in solid form.

Veneering Hammer: Tool used when gluing veneers.

Veneer Punch: A punch for veneer repairs.

Veiner: A tiny deep U-shaped gouge for texturing and veining.

Warp: The twisting of timber in drying.

Whittling: The craft of hand carving using a variety of knives.

Winding: A board with a twist is said to be in winding.

Winding Strips: Strips of accurately-planed timber used by the woodcraftsman when planing to test for winding.

Woodcarver's Screw: Used to hold blocks firmly to the bench top during carving.

Yankee Screwdriver: A screwdriver having a spiral ratchet for easy screwing.

Zig Zag: A type of folding rule, popular in North America.

Index

Entries in italics indicate illustrations

Bibliography

ADAMS, J T & STIERI, E: *The Complete Woodworking Handbook*, New York 1976
CONSUMER GUIDE: *The Tool Catalog*, New York 1979
CUNNINGHAM, B & HOLLINS, W F: *Woodshop Tool Maintenance*, London 1956
DUNBAR, M: *Antique Woodworking Tools*, New York 1977
GOODMAN, W L: *History of Woodworking Tools*, London 1964
HAMPTON, C W & CLIFFORD, E: *Planecraft*, Sheffield 1978

HASLUCK, P N: *Manual of Traditional Woodcarving*, London 1977
HOBBS, H J: *Veneering Simplified*, New York 1976
JACKSON, A & DAY, D: *The Complete Book of Tools*, London 1978
JOYCE, E: *The Encyclopedia of Furniture Making*, New York 1971
KRENOV, J: *The Fine Art of Cabinetmaking*, London 1977
McDONNELL, L P: *The Use of Hand Woodworking Tools*, New York 1978
MEILACH, D Z: *Creating Small Wood Objects*, New York 1978

MERCER, H C: *Ancient Carpenter's Tools*, New York 1975
NISH, D L: *Creative Woodturning*, New York 1976
ROBERTS, K D: *Wooden Planes in 19th Century America*, Fitzwilliam, NH 1978
SAINSBURY, J: *The Craft of Woodturning*, New York 1979
SALAMAN, R A: *Dictionary of Tools*, London 1975
SCHARFF, R: *The Complete Book of Home Workshop Tools*, New York 1979
SODERBERG, G A: *Finishing Technology*, Illinois 1969

Acknowledgements

6 Garrett Wade Company; 8 Mary Evans Picture Library; 10–15 Mike Fear, from the collection of Richard Maude; 17 Mary Evans Picture Library, The Mechanick's Workbench, Marion, Massachusetts; 18 Garrett Wade Company; 20 Record Ridgway Tools Ltd; 21 Helen Downton; 22–23 Garrett Wade Company, Record Ridgway Tools Ltd; 24 Record Ridgway Tools Ltd; 26 Garrett Wade Company; 28 Record Ridgway Tools Ltd; 30–31 Garrett Wade Company; 32–33 Record Ridgway Tools Ltd; 34 Garrett Wade Company; 36 Garrett Wade Company, Helen Downton; 38 Garrett Wade Company; 40 Helen Downton; 42 Helen Downton; 42–43 Garrett Wade Company; 44–45 Helen Downton, Garrett Wade Company; 46 Garrett Wade Company; 48–49 Helen Downton, Garrett Wade Company; 50–51 James Neill Ltd, Garrett Wade Company; 52 Garrett Wade Company; 54–55 James Neill Ltd, Garrett Wade Company; 56 Garrett Wade Company; 58–59 Helen Downton; 60 Garrett Wade Company; 62–63 Helen Downton; 64 Helen Downton; 65 Garrett Wade Company; 66–67 Garrett Wade Company, Helen Downton; 68 Garrett Wade Company; 69 Helen Downton; 70 Primus, Record Ridgway Tools Ltd; 71 Record Ridgway Tools Ltd; 72 Stanley Tools Ltd, Garrett Wade Company; 73 Garrett Wade Company, Record Ridgway Tools Ltd; 74 Garrett Wade Company, Record Ridgway Tools Ltd; 75 Kai Choi; 76 Stanley Tools Ltd, Helen Downton, Mike Fear; 77 Record Ridgway Tools Ltd, Stanley Tools Ltd, Mike Fear, from the collection of Richard Maude; 78–79 Garrett Wade Company, Helen Downton; 80 Record Ridgway Tools Ltd, Garrett Wade Company; 81 Mike Fear; 82 Helen Downton; 83 Record Ridgway Tools Ltd, Helen Downton; 84 Garrett Wade Company; 86 Record Ridgway Tools Ltd, Helen Downton; 87 Helen Downton; 88 Helen Downton; 89 Garrett Wade Company, Record Ridgway Tools Ltd; 90 Helen Downton, Record Ridgway Tools Ltd; 91 Garrett Wade Company; 92 Garrett Wade Company; 94 Helen Downton; 95 Record Ridgway Tools Ltd; 96 Garrett Wade Company; 98 Stanley Tools Ltd; 100 Helen Downton; 101 Firth Brown Tools Ltd; 102–103 Stanley Tools Ltd; 104 Helen Downton; 106 Garrett Wade Company; 108–109 Helen Downton; 110 The Mansell Collection; 111 Coronet; 112 John Sainsbury; 113 Craft Supplies; 114 Garrett Wade Company, Helen Downton; 115 Garrett Wade Company, Helen Downton; 116 Record Ridgway Tools Ltd; 118–119 Helen Downton, Garrett Wade Company; 120 Record Ridgway Tools Ltd; 121–122 Garrett Wade Company; 124–125 Helen Downton; 126–127 Garrett Wade Company; 128–129 Record Ridgway Tools Ltd; 130 Garrett Wade Company, Helen Downton, Record Ridgway Tools Ltd; 131 Garrett Wade Company, Helen Downton, Record Ridgway Tools Ltd; 132 Helen Downton; 133 Helen Downton, Record Ridgway Tools Ltd; 134 Garrett Wade Company; 136 Mike Fear; 137 Helen Downton; 138 Mike Fear; 139 Trevor Wood; 140–141 Helen Downton, Mike Fear; 142 Stanley Tools Ltd; 144 Helen Downton; 145 Record Ridgway Tools Ltd, Helen Downton; 146–147 Garrett Wade Company; 148–149 Helen Downton; 150 Helen Downton, Garrett Wade Company, Wolf Electric Tools Ltd, Elu Machinery Ltd; 152 Helen Downton; 153 Record Ridgway Tools Ltd; 154–155 Mike Fear; 156 Helen Downton; 158 Stanley Tools Ltd; 160 Record Ridgway Tools Ltd; 161 Helen Downton; 162 Helen Downton; 163 Helen Downton; 164 Stanley Tools Ltd; 165 Garrett Wade Company; 166–167 Helen Downton; 168 Garrett Wade Company; 170–171 Bosch; 172 Record Ridgway Tools Ltd, Wolf Electric Tools Ltd; 173 Skil Corporation; 174 AEG, Wolf Electric Tools Ltd; 175 Helen Downton; 176 Skil Corporation; 177 Rockwell International, Wolf Electric Tools Ltd; 178 Helen Downton, Skil Corporation; 179 Wolf Electric Tools Ltd.